CUSTOMER SERVICE
AND
HOTEL MANAGEMENT

CUSTOMER SERVICE AND HOTELE MANAGEMENT

M. C. METTI

ANMOL PUBLICATIONS PVT. LTD.
NEW DELHI - 110 002 (INDIA)

ANMOL PUBLICATIONS PVT. LTD.

H.O.: 4374/4B, Ansari Road, Darya Ganj,
New Delhi-110 002 (India)
Ph.: 23278000, 23261597

B.O.: No. 1015, Ist Main Road, BSK IIIrd Stage
IIIrd Phase, IIIrd Block
Bangalore - 560 085 (India)
Visit us at: www.anmolpublications.com

Customer Service and Hotel Management
© 2008
ISBN 978-81-261-3248-5

PRINTED IN INDIA

Printed at Mehra Offset Press, Delhi.

Contents

Preface

Delivery of quality customer service remains as important to the success of the hospitality industry as ever. The critical importance of fundamentals is not readily evident to the casual observer. But many a business failure appears to have occurred because the firm ignored the fundamentals of business. What is fundamental to any hotel business? To answer this question is to understand why a hospitality business exists and what it takes to deliver on it. There is only one valid definition of hospitality business purpose: to create a customer. The essence of that statement might seem simplistic, but then all fundamentals are indeed simple. That, in fact, may be the reason it gets ignored.

Perhaps its apparent triviality allows us to gloss over the fundamentals or take them for granted. Customer-orientation or customercentricity is not a new concept or practice and neither is customer relationship management (CRM). What is new is that information technology has unleashed tremendous opportunities in dealing with a customer and in creating value to the customer. It is ever more important that in leveraging these new technologies firms respect the fundamentals. Indeed, it is precisely because of the unprecedented attention that, as providers and consumers, we have given information technology that there has been ample opportunity to lose sight of the basic purpose of any business.

It is now becoming clearly apparent that all hotels compete on service. Service is offered as an inevitable and primary differentiation strategy — providing superior service becomes a prerequisite for any differentiation strategy to succeed. To provide superior service for a competitive advantage requires a concrete understanding of what service orientation means. This orientation in the form of a frame of mind is essential for the firm to take advantage of its opportunities and to address its challenges so as to gain a competitive advantage. For excellent service hotels, the challenges and opportunities in providing services are a constant presence. For others, these challenges and opportunities are less obvious.

However, that doesn't mean we have been doing a very good job at it. Over the last fifteen years, some customer service progress in hotel industry, but it has been slow in coming and spotty in impact. Unfortunately, the fact remains, as an industry, we are doing a very good job delivering a very mediocre customer service product. When it comes to customer service, mediocrity does indeed abound. Of course, there are horror stories that come to light from time to time, and sometimes, albeit far too rarely, hospitality. Customers receive a level of service that they can actually rave about. But overall, much customer service progress needs to occur before we can claim, as an industry, that we do in fact deliver a quality customer service product.

The lasting legacy of customer service in our industry lies in the hands of current and future leaders. Thus, the entire book has been designed to facilitate a comprehensive customer service review of any hospitality operation from a workplace point of view or an academic perspective.

Author

Chapter 1

Introduction

CUSTOMER-FOCUSED MANAGEMENT

Businesses fail when the fundamental question — what business are we in? — is either ignored or misunderstood. To answer this question appropriately, businesses have to be customer focused. What does being customer focused mean? Why is it that customer-focused firms achieve a sustainable competitive advantage? What does customer-focused management involve?

A January 2001 issue of the *Wall Street Journal* featured an article titled: "Can these dot-coms be saved?" The dot-com malaise was becoming widespread. The message in the article was simple, the question straightforward, but the answers were not readily evident. A closer look reveals that the answers are implicit in the questions within the article. "Is Priceline's vaunted name-your-own-price model good for anything but plane tickets?" "If traditional catalog retailers can make money selling clothing why can't a Web site like Bluefly?" "Can Webvan's net-fueled automation turn profits delivering cornflakes?" All three of these Internet examples represent a very typical situation that most if not all dot-coms, including that mainstay, Yahoo!, had been facing. The problem was endemic to the dot-com business. The technology was powerful, but in each case, the appropriate revenue model remained elusive.

The question on everyone's mind was: Where are the profits? And, when the profits arrive, would they be sustainable? The answer is in the question. If you asked the right questions, you could find the right answers. A Francis Bacon quote comes to mind: "A prudent question is one-half of wisdom." As the dot-com frenzy reached epidemic proportions, the right questions were not being asked, and the right answers were not available. All this may (yet again) seem simplistic, but many of the dot-com failures were at least partly attributable to a lack of due diligence. The analyses about sustainable profitability must be preceded by the fundamental question of what business the firm is in. Unfortunately, the times were such that the promises of information technology overshadowed the fundamentals and good judgment on the part of many among us.

THE BUSINESS YOU ARE IN

To make a profit in any business, you need to be able to sell for more than it costs you to produce. This is a basic prerequisite for any business, really. A viable business must produce enough revenues to cover costs and have surplus as profit. Ironically, the dot-com phenomenon brought the term "business or revenue model" to common parlance. The business or revenue model is a model that depicts how a business will generate revenue and what it will cost to do that.

The more important question may be, is the model sustainable in the long term? Sustainable profits come from sustainable competitive advantage, which in turn comes from consistently creating and delivering superior value. The concept of that superior value is essentially the business of your business. To really understand how you are going to make a profit, you have to first examine a more fundamental question: *"What business are you in?"*

- Priceline.com started with buying unsold airline seats at a heavy discount and allowing passengers to bid for them online. This is still its core business. Priceline experimented with selling gasoline and groceries and had to give these up. So, what business is priceline.com in? General retailer or broker of airline seats? Airline seats are perishable in that, if the seats are not sold before the plane takes off, the assets in the form of unsold seats perish. Priceline.com auctioning these seats off is a win-win situation for airlines as well as for passengers who are looking for cheap tickets and are willing to make last-minute travel plans. Why is it a "win" for airlines, and why is it a "win" for which type of customer? If this is priceline's business model, does it translate to gasoline and groceries?

- Can Bluefly compete with traditional catalog clothing retailers on price? Is Bluefly really in a different business? If not, it is in direct competition with established catalog retailers such as L.L. Bean and Land's End. Can these catalog retailers not offer a Web site retail front just as easily as Bluefly? Can Bluefly provide superior value compared to these established retailers, at a sustainable profit? If Bluefly is in a different business, what is it?

- Web grocers like Webvan are out of business. Why? Most likely, their analysis of the business model was flawed. Are they in the grocery business or in the delivery business? Can they sell what they produce/provide for more than what it costs them to produce it? And can they deliver better customer value than the competition? Who or what is the competition? Ultimately, the question is: Can they provide sustainable superior customer value at sustainable profit?

With very few exceptions, such as the Internet auction firm eBay, the job-hunting site Monster.com, and the online bank NetBank, just about every dot-com firm was not profitable when they issued their IPOs in the mid to late 1990s. In 2001, Priceline.com turned its first profitable quarter. Eight months after that article in the *Wall Street Journal* about saving the dot-coms, a more direct article in the same business-oriented newspaper provided a peek at the real cause of dot-com troubles: "Latest dot-com fad is a bit old-fashioned: It's called profitability."

A handful of dot-com firms including online travel sites Expedia and Travelocity, online brokerage Ameritrade, real estate listing site Homestore.com, and online ticketing site Ticketmaster were beginning to show promise of steady profits. Most Internet firms struggled with the fundamental questions or, worse, had not even dealt with them. Many relied on advertising dollars to support profits. In the mid to late 1990s about 600 dot-coms, including such short-lived but well-known Internet brand names such as Webvan and eToys, failed. The business models at many firms appear to have been flawed.

Many of them offered free content and banked on revenues from advertising on their Web sites. Facing severe drops in advertising revenues in a slow economy, most dot-coms began charging for content in a desperate bid to attain profitability. The focus had reverted to the fundamentals of doing business: What customer value are you creating and can you make a profit doing it?

Barry Diller of USA Networks, who has been successful in lining up cash-producing Internet firms, put it bluntly: "You have to deliver a better product to your customer, not just offer an Internet retail option for the sake of dabbling in some new technology." Better product than what? Better than previously available and better than the competition. Mr. Diller was actually referring to superior customer value. The experienced business leader never loses sight of the fundamentals.

Peter Crist, vice chairman of the executive recruiting firm Korn/Ferry International, should know about management talent. He was quoted in the *Wall Street Journal* as saying, "There's a whole generation of people in their 50s and 60s, who started a business and focused on the fundamentals that we're losing." Case in point is Heartland Express CEO Russell Gerdin, who spends a lot of time on details — details of the costs and revenues of his operations, route-by-route, on a day-today basis. Gerdin never sways from his focus: short-haul business, where he has a competitive advantage over the "big guys."

How firms go about determining what it takes to provide superior customer value at a sustainable profit is the fundamental issue for any business. If you miss the fundamentals, you will fail in just about any endeavor, in the business world or in any other walk of life. This book is just as much about fundamentals as it is about obtaining and sustaining a competitive advantage. You cannot attain that sustainable competitive advantage unless you have mastered the fundamentals of creating and delivering superior customer value.

As discussed, the analyses confronting the fundamental question of "what business are we in?" must begin with a clear understanding of who the customer is and what customer value is being provided. Is this customer value superior to the customer's alternative? Firms must ensure that they can provide this customer value in the long term at a reasonable profit. To be able to *sustain* the profits, the firm needs to provide this customer value better than the competition. Continuously providing superior customer value leads to sustainable profits from a sustainable competitive advantage.

To make *sustainable* profits, firms must be able to continue to provide customer value at a cost lower than what the customer is willing to pay. What the customer is willing to pay is commensurate with the perceived value of the solution. If factors of production are equally accessible to all firms, what makes perceived value of one better than another?

The answer, clearly, must lie in something that is not available to all. What is unique to a firm may very well be the strength of its customer focus. When firms are focused on the customer, they understand what customer value is. A clear understanding of customer value gives the firm a clear view of its competitive status among the customer's choice of solutions.

Delivering Superior Value Requires a Customer Focus

Every firm provides a bundle of values as a product. The firm that provides the superior value bundle wins the customer. *Superior* is a key word here. Why? Consider this. Research by Jones and Sasser has shown that even satisfied customers defect. These Harvard researchers were studying the link between customer satisfaction and customer loyalty. Their findings suggest that customers are satisfied at different levels. Different levels of satisfaction result in different levels of loyalty, which in turn result in varying levels of behavioural disposition to patronage with a provider.

This logic implies that there is a range of customer satisfaction levels. Satisfied customers may not necessarily translate to loyal customers. Other researchers have introduced the notion of delighting the customer. A customer may defect to a competitor even if satisfied with the current provider because the competitor may be offering a value bundle with the perceived potential for a *higher degree* of satisfaction for that customer. Therefore, firms have to strive to achieve higher levels of satisfaction than their competition by providing superior customer value. This is the essence of achieving a sustainable competitive advantage. If you are unable to provide superior customer value, even your satisfied customers could leave you!

What can a firm do that its competitors cannot match? To answer this question, one must clearly understand the context of competition — the landscape or playing field, as it was. Pine and Gilmore might call such a competitive landscape the "experience economy, " where products are quickly commoditized and firms compete on other aspects of the total offering. People, processes, and technology combine to create products. Technology is easy to copy, but customer-oriented employees and corporate culture are much harder to copy. In fact, technology is eventually commonly available to any firm that has the necessary capital and resources. Commenting on the recovery prospects of information technology firms after the recent recession, Marc Andreassen of Netscape says that a lot of innovation is going on now. However, most innovations are quickly becoming commoditized. One reason may be that processes and systems can be designed appropriately to deliver customer satisfaction, but they can be relatively easily replicated. Employee attitudes, however, are less easily replicable.

Customers look for convenience, cost, and quality of the total experience. The service-oriented firm concentrates on the capabilities of employees and not just on the technology or the tangible features of the product. The realization that the "clicks" store is not going to replace the "bricks" store has demonstrated that technology is not the solution. Rather, it is a means to enhance the solution. In this case, technology has simply provided a different solution.

Almost all bricks-and-mortar firms now have an Internet presence; it is an inexpensive service enhancement for them compared to what the dot-com firms have to do to establish a brick-and-mortar presence. Those dot-coms that have conducted the appropriate analysis of what customer value they were creating and delivering and how they were going to be superior to existing alternatives have had a good start. Subsequently, those that figured out how to sustain it while realizing the profit potential are still around.

It is not a stretch to say that many of the dot-com failures may have occurred because they were focused on the technology and not on the customer. Meanwhile, specialty retailers who are experienced at serving the customer have combined technology with their reputed customer service to make a winning combination. With the established catalog retailers, shoppers get an audio connection in real time over the Internet with customer service personnel who help customers with any questions they may have.

Experience the L.L. Bean or Land's End service at their Web sites to understand how well these specialty catalog retailers have adopted the technology of the Internet. Customer experience with these firms with or without the technology sets them apart from the rest of the players. Technology alone will not provide a competitive advantage. The combination of superior technology and superior employees with a service orientation will contribute to achieving sustainable competitive advantage.

You need a customer-focused and service-oriented culture within the firm. It also quickly becomes evident that you need suppliers as well as distributors and agents who have a similar and compatible culture. Thus, there are two strong themes in any discussion of "creating superior customer value at a sustainable profit": a customer-focus imperative and a service orientation. If all aspects of the firm and how the firm creates and delivers value to the customer are focused on the customer, the firm can provide a superior level of customer value. Such a firm is well positioned for a sustainable competitive advantage. Now, what does it take to be a customer-focused firm? What does customer focus really mean? And, what exactly is a service orientation?

Customer Focus and Service Orientation

A customer focus necessitates a deep understanding of customers and their activities, interests, and opinions around the particular value or solution that the firm is providing. It should be an attitude that is pervasive and that permeates throughout the firm such that it becomes ingrained as a culture. Once this focus becomes a given, then the firm will find itself in the mode of serving the customer while ensuring a reasonable profit. The realization is that you cannot achieve a sustainable competitive advantage to command sustainable profits unless you are customer focused.

Serving the customer contrasts with the notion of marketing a product. It is more than providing a solution for the customer. It is about *serving* the customer. The firm looks at the customer's need in the broadest possible context, going beyond the scope of the core product that satisfies the core need in a particular consumption activity. The product is redefined to include all value-added components in a total solution. By definition, the product is extended to include several dimensions that would include as part of the total product the value added in any and all customer interactions. The service-oriented firm is one that focuses on serving the customer, regardless of whether its core product is a physical good or an intangible.

Such firms are better positioned to deliver that customer value in its products. When all competing firms provide the same core product, the competitiveness of a firm or its superiority has to come from enhancements to the core product. The key to competitiveness is in these enhancements, which are invariably provided by service dimensions. These service dimensions contribute to providing superior customer value. To successfully provide these service dimensions firms have to be service oriented. When you look around, you will find that successful firms exhibit a frame of mind that is service oriented and focused on the customer.

The service-oriented firm owns responsibilities over and above just providing a product. Such a firm is proactive in anticipating customer needs and situations in customer interactions. For example, customer education is seen as a significant responsibility of the firm. The customer-focused firm views the customer's role in consumption activities as an integral part of its solution. A customer-focused firm truly cares for the customer. Such a firm's behaviour demonstrates that it feels privileged to have the customer. When its product fails the customer, the service-oriented firm will evoke built-in customer recovery procedures implemented by customer-focused employees.

The total customer experience at such firms goes beyond the core product and is superior to other firms. The quality of the total experience includes a whole host of components or experiences other than that of the core product. The value that the customer seeks includes all of these components and not just the core product. When Lexus, Toyota's luxury division, found that its customers in the United States were having to drive hundreds of miles to buy a Lexus, the carmaker converted trucks into mobile service stations so that customers didn't have to drive that far to have the car serviced. Singapore

Airlines provides its fliers a wide range of choices from what and when they eat to how and when they are entertained. The costs are worth the value the flexibility adds to the customer experience. Customers choose firms that provide this customer value in a total solution. Such firms are service oriented.

Firms add a number of dimensions typically encompassed in "services" to augment the product. Customer service is a common term used to denote these dimensions. But it is important to understand that customer service is a specific type of customer interaction and as such, it performs a narrow but necessary function. While customer service is typically involved with nonroutine customer situations such as recovering from a service failure, being "service oriented" can include, for example, anticipating that service failure and proactively initiating a service recovery. EMC, the storage devices provider, builds its products with redundant systems.

When the primary system fails and the redundant system takes over, the design and service team assigned to the client looks into the causes and alerts the client if necessary. EMC's diagnostics cover competitor products that are part of the storage and retrieval system. In service-oriented firms such as EMC, service recovery processes are built into the design of the total product offering. EMC's service orientation is reflected in the design of its products and in its concern for the customer's experience with the product performance. EMC claims a 99 percent customer retention rate. The total product offering includes serving the customer during all stages of consumption of a solution.

In service-oriented firms, services complete the product as a total solution. These services are also the primary key to a sustainable competitive advantage within that total value bundle. Providing these services requires a completely different mind-set with an understanding of the complex characteristics of services. This is true even in products where the core benefit is delivered by a physical good that the customer takes title to. The services that go along with the physical good are a significant component of the total product.

As illustrated in the EMC example, this perspective is readily apparent in the business-to-business (B2B) situation, where firms are naturally inclined to focus on service to the customer. Their perspective includes service elements to augment the product and to establish and maintain customer relationships. For manufacturing firms, the physical good is only part of the total solution toward customer needs. Indeed, the physical components of any product are easily matched by competitors, thus reducing what might have been a competitive advantage to a commodity. Firms really compete on the noncommoditized part of the total product. When these aspects of the product are meaningful to the customer and the firm can provide customer value better than the competition, the firm has a sustainable competitive advantage. Firms compete with each other on the total solution comprised of commodities and noncommodities — a complex set of benefits or value bundle.

For a number of firms, especially in the information technology industry, the product has become the service more than the physical technology itself. By the year 2000, for example, 40 percent of sales and 40 percent of profits at IBM came from its Global Services division. At Unisys, 69 percent of sales came from services. Sun's $3.3 billion services business was closely tied to its server (the black box) business. Compaq purchased DEC in 1998 for its service business. Hewlett-Packard later purchased Compaq for the same reason.

A number of articles in the popular press have lamented the sorry state of service in the United States. When packaged-goods firms are included in the list of examples, it is in the service component of the total product that the failures are apparent. Why it that firms are unable to provide the desired levels of service to satisfy customers?

Beginning in the late 1980s, there has been a great deal of research on what exactly customer orientation or "customer focuses" means and on what exactly a firm should be doing to implement such an orientation. Peppers and Rogers, known for their one-on-one marketing concept, consider customer-focused management as synonymous with relationship marketing or customer-relationship management. Information technology offered firms a way to obtain, process, and use individual customer information so that firms would be able to personalize customer experiences. Some scholars have called it "market orientation."

One set of researchers suggested that customer orientation is a subset of market orientation. Their definition for *market orientation* is "the set of cross-functional processes and activities directed at creating and satisfying customers through continuous needs-assessment." Following their lead, market orientation has been treated as being composed of three components: customer orientation, competitor orientation, and interfunctional coordination.

Can the terms *customer orientation* and *market orientation* be used interchangeably? If we accept the definition of a "market" as being a set of potential customers and treat the terms *market* and *customer* as synonymous except for the level of aggregation in numbers, then we can use the terms interchangeably. Such an argument does not necessarily negate the three-component structure of the concept of customer focus. To be truly customer focused, the firm has to be driven by the goal of providing the customer with the highest level of satisfaction. This implies that the firm concentrates on how the customer is better served (by the firm) compared to the competitive offerings and that all processes and activities in the entire firm are integrated and coordinated to accomplish this goal. The customer-focused firm has all aspects of the firm directed at providing the superior product to the customer.

Thus, we can define *customer focus* as:

A form of culture in a firm that directs all processes and activities of the firm toward providing superior value to the customer so as to sustain long-term profits.

It is now generally accepted that market orientation or a customer focus is a type of culture and is exhibited by a firm that is committed to providing superior customer value. When compared with different types of business cultures, the customer-focused culture has been found to be superior in delivering the best business performance. Such a customer focus was strongly correlated with the use of customer information in a study involving about 5,000 salespeople and sales managers. This study also found a strong correlation between market orientation, job satisfaction, and trust in management.

Another study of 278 salespeople and sales managers found that market orientation significantly influenced job attitudes and the customer orientation of the salespeople. A firm was seen as clearly supporting the salespeople when the salespeople perceived that the firm was being attentive to customer needs and satisfaction, aware of competitor strategies to deliver superior value, and coordinated through the entire firm. There are proven bottom-line benefits from a customer orientation. In another study using a sample of 127 strategic business units in firms listed in the Fortune 500, customer-focused firms achieved higher profitability, sales growth, and new product success.

With such evidence of strong correlations between customer-focused cultures and successful business performance, it behooves us to learn what these firms are doing to be customer focused and what it takes to be a customer-focused firm. It is clear that there are firms that have been successful in developing and implementing the customer-focused culture. Scandinavian Air Systems, Walt Disney, and British Airways are firms that pay attention to establishing the appropriate culture and climate that fosters customer orientation. The list of issues that have to be addressed runs the gamut from organizational structure and behaviour, operations, and value delivery mechanisms to internal and external metrics such as in balanced scorecards and customer loyalty measurement. This book will cover many of these issues in a comprehensive yet concise manner.

Based on the critical links in the logic espousing the merits of a customer-focused management, we first look at the elaboration of the fundamental premise of customer value and service orientation. Following this we begin to examine the specific issues in developing a customer-focused and service-oriented firm — managing customer information, creating and delivering customer value, managing customer relationships, and ensuring a customer-oriented culture. Thus, the rest of the book is organized into the following five parts.

CUSTOMER VALUE AND SERVICE ORIENTATION

Understanding the business begins with a clear understanding of customer value. Providing a superior customer value requires an understanding of service orientation. Against this backdrop, the rest proceeds to follow the logic that sound decisions and actions cannot be made and performed without a solid understanding of the customer.

You cannot claim to be customer focused if you don't have the appropriate information on the customer. What kinds of customer data are important and why? How do we collect, interpret, and use this data? Where are the opportunities for gathering customer information? Information on all customer preferences, attitudes, and activities associated with the acquisition and use of the product provides a deeper understanding of the customer. This part develops the various methods and approaches to ensuring that the firm is adequately prepared to make customer-focused decisions.

Customer information should drive the design and delivery of customer value. What is involved in conceiving and designing a customer-focused product offering or service delivery? How would a firm go about tracking whether it is maximizing the returns from its value-creating assets? In this part are about effective and efficient value creation and delivery.

Who are the right customers and how can the firm ensure their loyalty to the firm and its products? This part covers topics such as service quality and guaranteeing quality service.

When the product fails the customer, what does the firm need to do to recover from that failure in the context of customer service? What must firms do to proactively manage customer defections?

How does the firm create and maintain a customer-focused culture — a culture that is so focused on the customer that all decisions and actions of all personnel and systems in place create superior value for the customer at a reasonable profit? This final part provides the framework to analyze and manage the firm's value systems to be customer focused.

The principal concepts can be stated as a truism: Businesses cannot succeed in the long term unless they are focused on its customers so as to provide them with superior service. This is how firms attain and maintain a competitive advantage and therefore provide superior customer value at sustainable profits. Thus, the two fundamental and closely related concepts evident here are "customer focus" and "service orientation." After reading this book, you will be able to engage yourself and the firm in a service-oriented mode of designing and delivering customer-focused value to maintain profitable relationships with the right customers. You will appreciate what is required to sustain a competitive advantage for sustainable profits by:

- Adopting a customer-focused and service-oriented frame of mind
- Understanding the business by understanding customer value
- Capitalizing on the opportunities and use of customer information
- Designing and delivering profitable customer value
- Focusing on the right customer and managing that relationship
- Establishing a customer-focused and service-oriented culture in the firm

Customer-focused firms are the product of customer-focused employees and processes. Customer-focused employees are the responsibility of the executives of the firm. As an executive, you will find that when you stop and think of the ideas and concepts, you will have begun to engage in the intellectual exercise leading to a customer-focused and service-oriented frame of mind. You will be able to specify what the firm should actually be doing to be customer focused.

CUSTOMER VALUE AND SERVICE ORIENTATION

Customers buy solutions, not products. Every firm needs to understand what it is providing the customer in terms of customer value, not by focusing on the product but by focusing on the customer. In doing so, the firm is able to assess whether it is providing superior customer value and whether it has a sustainable competitive advantage. Competitive advantage cannot be sustained unless it can be protected from competitor matching. Any aspect of the product can be matched by the competition. What is most difficult for the competition to match is how the firm approaches and treats the customer. The firm *serves* the customer by providing product enhancements in the form of service. When other firms can provide the same service, the differentiator is in the service orientation of the firm.

Customer Value

Many firms focus on the product rather than on the customer. When the focus is on the customer, the business is defined in terms of customer value, not in terms of product. The product is viewed as a customer solution and experience. That is, for any firm, the most fundamental definition of its business.

Try this one. What business is Amazon in? That you can buy just about anything on Amazon may not be too much of an exaggeration. The company abandoned or at least cut back on its strategy of buying up companies, as it did with Drugstore.com in 1999. It now handles retail for such established brands as Toys 'R' Us, Target, and Circuit City. Amazon shocked everyone in early 2002, when they announced their first operations profit of $59 million and a net profit of $5 million in the previous quarter. What value does Amazon provide its customer?

Customers don't buy products. They buy solutions. If firms are focused on the customer, they ought to see themselves in the role of providing solutions, not products. The information technology sector has popularized the term "solutions providers, " with labels such as application service providers (ASPs, as they are called) or technology solutions providers. These labels may be appropriate at the very general level and allow people to gloss over and take for granted the real meaning of the term *solutions*. The label has no real meaning as a concept unless it can be translated into customer value terms. Customers buy value in the solutions. Thus, customer value defines the primary purpose of any business.

This presents an approach to conducting a critical analysis of this fundamental question of "defining the business of a firm." In doing so, it defines "customer value" and redefines "products" in the customer-focused perspective. Finally, this briefly indicates the fruits of a customer-focused organization providing superior customer value.

Management guru Peter Drucker said in his 1954 book, *The Practice of Management*, "to know what a business is we have to start with its purpose." He goes on to say: "There is only one valid definition of business purpose: *to create a customer*." Drucker reiterates this concept in a number of different ways, always emphasizing that "it is the customer who determines what a business is." It is not important what the business thinks it is producing.

The customer determines the value of what the business produces based on what the customer thinks he or she is buying. Such a perspective places the final result in focus — what the product *does* for the customer. Thus, the purpose of a business is what it creates for the customer. The first and foremost question for any firm to ask is, what business are we in? How many firms diligently confront this question? And, when they do encounter this question, whose perspective do they take?

Taking the customer's perspective to answer this fundamental question forces a number of thought processes that benefit the firm in different ways. In fact, businesses will find that approaching the question of *what business are we in?* in so fundamental a manner not only allows the firm to understand its existing customers, but will also help open up new market potential. It forces the analysis of value-creating assets, including the firm's skills and knowledge sets. The question also opens up the analysis of who the competition is. When firms strive to beat rivals, they typically end up competing within the confines of existing business.

Firms sometimes do not realize that they compete with firms in other industries. They are able to see this if they articulate their business from the customer's perspective. When customers consider their options in meeting a need, they don't limit themselves to one industry. Without realizing it as such, their solutions may involve different industries. Firms are forced to redefine their competitive landscape when they take the customer's perspective because they will find themselves competing outside their well-defined spaces. Kim and Mauborgne call it competing in a "market space.

They urge firms to expand their view of their business by looking "across substitute industries, across strategic groups, across buyer groups, across complementary product and service offerings, across the functional-emotional orientation of an industry and even across time." They argue this point with examples of how Home Depot cultivated the do-it-yourself market from homeowners using contractors, how Intuit saw that people managing personal finances using a pencil and paper calculation actually competed with their software, how overnight package delivery firms FedEx and UPS competed with telephones and fax machines, and how Southwest Airlines competed with driving. This argument illustrates the definition of a business from the perspective of the customer. Not only does it underscore the value that the customer places on what the firm delivers, but it also reveals that the customer's perspective redefines the business.

Well-known Harvard scholar Ted Levitt expanded on Drucker's thesis of defining a business from a customer's perspective when he popularized the concept "marketing myopia, " where firms and entire industries (such as buggy-whip makers) could perish when inwardly focused. He used the example of railroads as an industry that suffered from marketing myopia because it did not see itself in the business of transportation, but rather in the railroad business. Contrast this with a firm like the Williams Companies, one of the world's largest pipeline builders. Until a few years ago, it built steel pipes to move oil and natural gas. Today, it also builds fiber-optic pipes for the big cable companies. Clearly, Williams did not see itself as being in the oil and gas business, but as being in the business of serving big companies that needed to move material (or data) through pipes. Would Williams have seen this opportunity if it had been focused on competing within the oil and gas industry?

To expand your opportunities, you need to look beyond the product horizon and into the customer's space. For example, if a family was looking for something to do on a Saturday, consider their options: have some friends over for dinner, go out for dinner, go to a nearby mall, go to the movies, go to a play or a show, go bowling, go to a miniature golf course, go for a drive to the lake or the beach—and many other options. Now, look at these options in terms of all of the different industries that are competing with each other. If you were a firm in any one of these industries, you could be defining your business the same way as any of the other industries, depending on who your typical customer is. The customer's perspective broadens your own view of your product.

Taking the customer's perspective helps firms identify what they are about—their mission and their values. They are better able to understand their strengths and sources of competitive advantage. Firms are better positioned and more effective in allocating resources and efforts, both strategically and operationally. They are better situated to understand what would be considered superior customer value.

Products as Solutions

Where do you even begin when you want to define your business? Let us consider the basics. What *do* businesses sell, at the most general level? In what terms will your response be made? Do you see services and goods as solutions to a customer problem—a need to be met? Customer needs are problems searching for solutions and firms are providers of solutions to customer problems. When seen this way, firms draw focus away from products and orient themselves to the customer with the sole purpose of delivering solutions to problems. The business is defined in terms of customer solutions. Firms suffering from a product focus define their product by its capabilities in terms of product features and promptly lose sight of the customer. The solutions-to-problems approach emphasizes the customer rather than the product. It emphasizes product benefits rather than product features.

Firms must align processes, people, and their entire culture to serving the customer rather than on making and selling products. In customer-focused firms, processes are designed with customers in mind. The output of the processes, the product, is defined by the customer need it serves. In serving the customer, firms offer a combination of benefits or values. Are all products, therefore, a service to the customer? In a sense, yes! The solution represents the value that the firm is providing customers by serving their needs.

Firms' activities produce solutions as value bundles that we call products. All products involve a mix of value producing components—tangibles and intangibles—in value bundles. The proportion of tangibles and intangibles in the entity that the customer takes title to prompts the artificial distinction between services and products. This unfortunate dichotomy detracts from the necessary customer focus in the conception of the product. Regardless of whether the total product is predominantly a packaged good or a service, in a business-to-business (B2B) or business-to-consumers (B2C) context, all solutions to customer problems can be seen as bundles of benefits or value bundles designed to deliver value by meeting customer problems or needs. This book adopts the approach that all products are a mix of physical goods and services in some proportion.

Some solutions employ more tangibles (physical goods) than intangibles (services) in meeting customer needs, while others may provide an intangible (service) involving physical objects. The important question is, what is that total solution from the customer's perspective? It is more important to understand what the value bundle *does* for the customer than what the bundle *is*. To identify the value bundle, one needs to identify and understand the customer.

Identifying the Customer

To understand the customer is to understand your business. To understand the customer's perspective is basic to defining a business. The first obvious question is, who is the customer? If the firm defines the customer too narrowly or without context, any subsequent analysis is flawed. Firms either focus on their end users as packaged-goods firms tend to do, or as in the case of industrial firms, the focus is on the immediate customer. Both cases reflect a narrow view and a lack of understanding of the context. To determine who the customer is, the question to ask is, how does what we produce or provide add value to the client's value creating process?

Most important, what do customers see as the value we are providing to their value creation process? This point is central to what we know as the "value chain" concept. An understanding of customers requires an understanding of the customers' value chain that includes the set of their value consumption activities. If the value created by the immediate customer creates value for a subsequent customer, that customer is also an indirect customer of the firm. We really need to identify the customer as a first step to determine whose perspective we should take and to understand the customer value that we are providing.

Performing the analysis of the customer sometimes forces the redefining of the customer. For Bright Horizons, a workplace child care and early education service company, it made a huge difference in how they went about their business when they viewed employers, and not parents, as their primary customers. The company could now tap into the financial and other resources of corporations and gain access to a much wider pool of parents. The value they were providing to the employer became the focus of the company. Chase Manhattan, for example, figured that its child care centre was yielding 110 percent return on investment through reduced absenteeism. Similarly, Merck found that its employee retention rate among those who were young parents went up dramatically.

The customer-focused perspective blurs the boundaries between entities in the value chain. Take the case of Aramark Corp., a firm that caters special events and runs cafeterias for big corporate clients. Nancy Naatz, resident district manager for business services for Aramark, has an office in the premises of her customer, Sears, in Chicago. Since Ms. Naatz is on location at her client's company, she is able to observe and interact with the local vendors Aramark has contracted to supply and serve Sears. Ms. Naatz is able to understand Sears executives' and employees' nutritional needs and interact with the vendors to ensure that the appropriate choices are made available. In this case, Aramark matches the food vendors' solution with the needs of the Sears employees who are the real customers.

So, who is the customer? The perspective of *which* customer I inspect and analyze, so as to help determine the business of my business, is the question. This question also raises the fundamental marketing decision of what or who our target market should be. It is a logical place to start before making any marketing or operations decision. Of all the potential customers that make up the market, which particular type of customer is most appropriate for our solution? The analysis involves segmentation, and the decision involved is one of targeting.

We are able then to determine which particular type of customer from the universe of all customers for this product would find most value in what our value-creating assets can produce. We thus segment the market and target those segments for whom our value-creating assets can provide differentiation and a cost advantage. To know who the customer is necessitates an understanding of the typical customer in that segment at a general or aggregate level.

Traditionally, customer profiles have come out of the segmentation exercise. Segmentation variables in the B2C context have included demographic, psychographic, and behavioural variables. For customer-focused management we need information on all of the characteristics of the need or problem or use-situation that the customer is involved in, encompassing all of the activities of the customer that the product is a part of. Consequently, the question, who is the customer? You want to know how the customer purchases and uses the product.

Similarly, the profile of the customer in a B2B context involves understanding the business of the organizational customer. For either the B2C or the B2B context, an understanding of the customers requires analysis of their use-situations, and the contexts within which a product helps them in what they do. Of what value are you to them? Now we begin to hint at the notion of what customer value really is. The answers lie in the *consumption* of the product, not in its production.

Customer Value

Customer value is what the customer thinks he or she is getting in return for what the customer has to part with, reflecting an implicit comparison akin to "give and receive." It has been described as the quotient of quality over price. Understanding the customer is about understanding customer value. Customer value is a complex concept and measuring it is complicated. But in the exercise of trying to understand customer value, the manager benefits from a better understanding of the customer and of the opportunities for a superior solution. Here are some characteristics of customer value that any manager should examine in an analysis of the firm and its activities: Customer value

- Is what the customer believes that a product or service provides in a certain use situation.

- Is an implicit comparison between what the customer receives from the provider and what the customer provides in time, effort, and money. The frame of reference is not just the price tag on the product, but also the ease and convenience in the acquisition and use of the product and the whole interactive experience the customer has with the firm.

- Is the customer's rather than the provider's perspective. It has to do with customer perceptions of the product and the firm, and not the provider's perspective of what the firm is delivering.

- Is dynamic, in that it can change over time before, during, and after the purchase, use, over repeated use of the product, and during all the various stages in the relationship with the providing firm.

- Can vary over different use situations. Your automobile may provide less or more value in transporting a group of kids to the ballpark compared to taking the spouse to the boss's house for a party.

- Can be shaped by attitudes, opinions, and behaviours of others, such as friends, family, media, the providing firm, and other competitors in the industry as well as in substitute industries.

- Determines customer satisfaction and the likelihood of brand or firm loyalty. This cause-effect relationship could be affected by comparisons with competitors and other substitutes.

- Is hard to measure and keep track of. Customers may find it hard to articulate all the different dimensions of customer value and firms may find it a rather onerous task to keep track of all the manifestations of customer value.

Your Value to the Customer

In order to determine what business a firm is in, therefore, the fundamental and imperative exercise is to understand customer value. Let us begin the analysis with the question, what exactly does customer value consist of? It is not easy to determine what the customer is getting as value from a product or service. To understand customer value, you really need to get into the customer's way of thinking about products. In a recent *Harvard Business Review* article, Chase and Dasu urge the use of behavioural science to get into the head of the customer, to understand their total experience with the product or service. Only then can you grasp all the components of customer value.

Examining the components and determinants of customer value is a critical step in the analysis. It requires analysis of the benefits as well as costs to the customer. The benefits seen are evident in the perceptions of the product performance. Perceptions of product performance are framed in the context of customer expectations. The expectations are shaped by past experience with the product and by messages about the product received from a variety of sources such as friends and family as well as from the marketer.

From an analytical perspective, product performance comes from functional (core) benefits and supplemental benefits. Functional benefits arise from the core attributes of the product. For instance, you buy a set of golf clubs. How the golf clubs affect your game is a core and functional benefit, as in the primary benefit from the product. The prestige of the brand image, the product return policy, and other customer service features would translate into supplemental benefits. Every customer value analysis requires that we identify and examine core and supplemental benefits.

Similarly, we need to identify the costs to the customer — both monetary and non-monetary costs. Monetary costs include costs incurred in the acquisition, use, and disposition of the solution. Non-monetary costs could include the time and effort in acquiring and benefiting from the solution and the opportunity costs where the product failed. Tom Wright, Buy.com's vice president of operations called his company's old way of handling returns "almost embarrassing." To return a product that a customer bought from Buy.com, the customer had to telephone Buy.com to generate a return authorization, which led to a shipping label from the package delivery firm (UPS) to be mailed to the customer, who then used that label to send the product back to Buy.com. It was weeks before the customer got credit for the product return! Imagine the non-monetary cost of aggravation to that customer.

Many firms that focus on acquiring customers do not retain customers because of their shallow understanding of customer value. Customer value includes interactions with the firm before, during, and after the consumption of the product. Smart firms are focusing on the whole customer and defining their business in terms of market spaces. The market space is the actual set of customer solutions defined by the consumption activities to which the firm caters.

In recognizing that customer value is dynamic, we need to place this analysis in the context of different use situations and how they are changed over time by various influences. Customers' concepts of what to expect from a provider change with each consumption experience. They learn from their own experiences with the product as well as vicariously from others. Competitor offerings and marketing messages continually shape their expectations. Therefore, implicit in the assessments made by the customer is a comparison with expectations based on the alternative — that is, the competition. Customer value analyses can be performed at a qualitative level and with some difficulty can be quantified as well.

Data-driven representations of monetary worth of what a firm does for a customer are what Anderson and Narus called "customer value models." Such models are useful in assessing the customer value of your product and comparing it with the competition to determine whether you are providing superior customer value or not. The next fundamental question is, Can you provide this superior customer value at a sustainable profit by commanding revenues to cover costs and profits?

Mutual Value of Customer and Firm

Just as the customer value analysis requires an examination of the benefits versus the costs, the firm makes a similar analysis to determine its profit potential. In providing the customer value, what does the firm get in return? How do the monetary and nonmonetary benefits from the customer compare with what it cost the firm to provide that solution to that customer? Can we sustain these profits in the long term? A firm creates value for itself by creating value for the customer. In many ways, the value of a firm is reflected in the profits it provides its investors and the value of the customer to the firm is reflected in the profits that the customer provides the firm. The greater the values of the customer to the firm, the greater are the profits.

This means that the surplus from the revenues after costs is greater if you can attract and retain profitable customers in the long run. Part IV in this book covers the valuation of the customer by the firm. Firms can attract and retain customers if they can provide superior value. It could boil down to, can we command a premium on our solution? If you can, it turns out that the margins are not in the commoditized part of the product, it is in the services provided by the firm. The margins increase when you are able to extend your role in the customer's space. This may come in the form of "cross-selling" or "up-selling." It may be the case that the total solution provides such value to the customer that the customer is better off with you than with the competition.

As the firm gets more valuable to the customer, the customer gets more valuable to the firm. The value from a customer is derived not only from purchases. Customers can become advocates for the firm, the benefits of which are frequently less understood. The value of the customer in terms of purchases and referrals over the lifetime of the customer should be an imperative calculation for firms who are focused on the customer. It appears that in the final analysis, it is the customer-focused firm that can provide a complete solution through a service orientation that succeeds in providing superior customer value and in return enjoys sustainable profits. So, how is it that a firm is able to provide superior customer value? What gives them a sustainable competitive advantage, such that the firm has sustainable profits?

Delivering Superior Customer Value

What characterizes firms that can provide *superior* customer value? To answer this question, we have ample evidence from research on such firms. Every firm in any industry provides some customer value. The firm that provides superior customer value is the one that has a sustainable competitive advantage. Firms compete with each other on differentiable aspects of the total solution. If we can conceptualize that every product has parts that are commodities and parts that differentiate it from others, superior value is in the augmented product, not in the conceptual core.

For example, all airlines provide transportation from point A to point B. But the total solution includes aspects other than the physical transportation of the person itself, such as amenities on the airplane. A firm must seek opportunities to differentiate itself from the competition in ways that are meaningful and relevant to the customer. How do firms find opportunities to differentiate themselves in a way that it is sustainable?

Consider how CDW creates superior customer value. CDW is the leading B2B e-commerce firm selling technology equipment to small and medium-sized firms. Wall Street analysts claim that the firm uses a "clicks and people" strategy. Supposedly, their competitive advantage is superior customer service and lower customer acquisition costs. The firm's Web site is tailored to each specific customer as a "custom extranet." How are they able to sustain this competitive advantage? Their own chief financial officer, Harry Harczak, notes that retaining people is key. In his experience, the productivity difference between an employee who has been at the firm six months versus one there three years is a factor of six. CDW values employee commitment.

What are the generalizable characteristics of firms that create superior customer value? A team of Harvard Business School researchers set out to discover exactly this and concluded that successful firms such as BancOne, Intuit, Southwest Airlines, ServiceMaster, USAA, Taco Bell, and MCI made employees and customers paramount. These researchers modeled the findings from their case studies into a framework they called "the service profit chain." Their cases provided data to establish powerful relationships between profitability, customer loyalty, and employee satisfaction, loyalty, and productivity. Evident in the whole model is the focus on the customer and the service orientation of satisfied employees, and the implications on investors and other stakeholders of the firm.

As firms focus on customer needs and become service oriented to deliver satisfaction, there are favourable ripple effects throughout the firm. Consistent delivery of customer satisfaction persuades customers to be loyal because perceived risk in buying a product from that firm is substantially reduced. When providers of solutions to customer needs are able to consistently deliver expected value, customers are more likely to be consistently satisfied. By way of their continued patronage, they become more valuable to the firm. When customers are loyal to the provider by choice, it is also in their best interest to support the provider because the prosperity and longevity of the provider is beneficial to the customer. Similarly, with regard to employees, research has shown that when employee satisfaction is high employees are more likely to be loyal to the company. When employee loyalty is high, employees are likely to be more productive and to be providing quality products and service.

Neither employees nor customers are likely to be loyal unless their expectations are met and they are satisfied.

Therefore, customer satisfaction and employee satisfaction are inherently tied to each other, in that, when employees are loyal to the firm, they provide superior performances driving customer loyalty. When employees are productive and customer loyalty is high, the firm is successful and can command a premium.

With increased profitability and growth prospects, investors are attracted to the firm, which in turn makes more resources available to the firm. With more resources available to the firm, capital is available to continuously improve value-creating assets (people and processes) and to produce consistently superior customer value in its products and service. This sequence of events, presented as the "loyalty circle", contributes to profitability and growth and requires that people and processes are aligned to deliver superior customer value. Delivering superior value should be a relentless endeavor for the firm.

THE CUSTOMER-FOCUSED ANALYSIS

The analytical frame to understand customer value consists first of identifying the customer and understanding the customer's problem and solution needs. This leads to an understanding of what the solution does for the customer and everything associated with it. The next issue is to focus on the options that the customer would consider appropriate in meeting the need and how the options compare with each other. Are there other options provided by other firms whose value-creating assets have a sustainable competitive advantage? This would dictate whether the firm should be in this business at all; if so, is there the potential for a sustainable competitive advantage? The kinds of questions that need to be asked by the customer-focused firm as it examines its business and how it can provide superior customer value for a sustainable competitive advantage.

The Key to a Sustainable Competitive Advantage

The core of any product is essentially a commodity. Competitive advantage comes from services as product enhancements made to the core. To compete on service, firms need to understand the opportunities and challenges presented by the special nature of services. Sustainable competitive advantage comes from an ability of the firm to compete through service.

A common misconception is to view services as being synonymous with customer service. "Services" are not limited to customer service. Customer service is only one of the services that enhance products. Managers who are product focused and not customer focused usually see services as meaning customer service. They see service as something that is done when a product fails or invoked to prevent it from failing. Others who see service as "freebies" that you toss in with the product are probably sales focused.

Firms with such a product or sales orientation actually do a disservice to the customer in the name of service. Their tendency is to offer a refund or return policy, which is often enough structured to favour the firm rather than the customer. These firms may offer a customer service toll-free telephone number, which is usually set up to be a frustrating encounter for the customer service personnel as well as the customer who may have been calling in for retribution. Such is often the unfortunate consequence in firms that are not customer focused. In customer-focused firms, customer service is not just an add-on, a little something extra to keep the customer happy. Customer service is an integral part of the total product.

Service is not what you *do to the product*, it is what you *do for the customer*. The customer-focused firm looks at the core product as an incomplete solution for customers. These firms supplement the core product with services that help customize or otherwise enhance solutions for customers. Such product enhancements add value for the customer. The Aramark employee being on location at Sears adds value to the core product of catering for Sears. It allows Aramark to be more customer focused by obtaining first-hand customer information and being able to deliver on it. By being on hand, Aramark can be immediately aware and responsive to the customers' needs. Service-based product enhancements are an imperative for the customer-focused firm.

This establishes the frame of mind required for the customer-focused firm to be service oriented. You will see what the inherent characteristics of services are, and thus the challenges and opportunities you need to recognize when you incorporate services into the value you create and deliver to the customer.

The Service Orientation Imperative

The traditional view of services is to treat products dichotomously, in two categories — *either* services *or* (physical) goods. Sometimes the term *product* is even used interchangeably with packaged goods but not with service. Not only has this usage caused unnecessary confusion, but it has also encouraged the flawed approach that services and packaged goods are separate and mutually exclusive entities. Scholars have gone as far as to suggest that "most product manufacturers and service providers alike are largely service operations."

They asserted that the role of services is critical for any organization in providing value in the form of "technological improvements, styling features, product image, and other attributes that only services can create." The fact is that all products come with some services. As a matter of degree some products have more services than others and are therefore more intangible than others.

Picture a continuum ranging from products that are most tangible (and least intangible in proportion) at one extreme to products that are most intangible (and least tangible in proportion) at the other extreme. At one end of the continuum, services such as education, consulting, and financial services have very few physical goods that customers take title to. At the other end, packaged goods such as a bar of soap or table salt come with no apparent service unless, for example, the customer initiates a customer service phone call.

In the middle of the continuum are products such as fast food restaurants and custom-made clothing that have an almost equal proportion, with no real predominance of tangibles or intangibles. In products that are predominantly intangible, where the customers don't take title to anything physical, such as in banking services, the service provider might use tangibles such as documentation, statements, and billing. The value for the customer is in the information contained in them.

As a proportion of what the customer is getting in the total product, the predominant source of customer value is from the intangibles in the total product. For physical goods, services enhance the core product and provide opportunities for competitive advantage. Thus, any product has some proportion of services attached to it, and this part of the product has a special nature that needs to be handled differently. If all firms provide some proportion of service components as part of the total product, it behooves them to have an understanding of the nature of services.

Only with such an understanding can we truly get into the service-orientation frame of mind. Take for example, the table salt manufacturer, Morton's. For this firm to look at itself as a packaged goods firm and therefore not concerned with services is a mistake. The fact that Morton's makes the product available at the retail store through the appropriate distribution channels is a service to the customer.

Morton's has essentially outsourced its distribution and retailing to channel intermediaries. To be truly customer focused, Morton's needs to view all the value-added it provides the customer, over and above the product of the production process—the salt—as services it provides to the customer. Morton's cannot manage its service components in the same way that it manages the production of its table salt. Most B2B manufacturing firms intuitively recognize the importance of the service component in enhancing its product. Caterpillar, the earth-moving equipment manufacturer, recognizes that prompt and reliable service is critical to its success and organizes the whole firm around customer locations. You need a service orientation to see such product enhancements as service dimensions that add value to the core product. Service orientation is

- A philosophy or frame of mind reflected in the firm's culture
- An attitude to serve the customer reflected in the firm's treatment of its customers, and
- A view of services as necessary enhancements to any core product to make a complete solution.

To provide complete and competitive solutions to the customer, one needs to understand and adopt the service orientation. The service-oriented frame of mind requires a grasp of the fundamental nature of services (whether as the product or a component of a product). Once the inherent characteristics of services are clear, it becomes apparent that their implications for the customer and the provider offer a number of opportunities and pose a variety of challenges.

As a preface to a discussion of these issues, it would be useful to note the evolution of interest in the concept of service. Prompted by the frustrations of practitioners in the service sector who were finding that marketing practices from the packaged goods world did not make sense for services, scholars began to get interested in the problem. In the late 1970s and early 1980s, for a number of reasons, the academic research community in business disciplines, particularly in marketing, operations, and human resources, began a serious intellectual debate as to whether products that were services, compared to products that were packaged goods, needed to be studied differently.

For example, could you study the marketing of hospitality services the same way as you would the marketing of toothpaste? Some scholars maintained that marketing is marketing regardless of *what* you are marketing. Either way, they argued, one had to go through the tasks of segmenting the market, positioning a product, and making product, pricing, distribution, or promotion decisions. Others argued that although that might be true, one would need to approach these tasks very differently; and they proceeded to offer the rationale for this argument. There was a great deal of interest in this effort, especially among those who recognized the tremendous growth of the services sector.

Around the same time there were environmental changes in the economy and in the competitive landscape. Services sectors such as airlines, telecommunications, and, later, banking and insurance were going through deregulation in the United States. Manufacturing was moving to cheaper labour markets in the Far East and in Latin America. Meanwhile, by the late 1980s in the United States, even professional services such as health care and legal services began to recognize the need to employ marketing practices.

Not surprisingly, even manufacturing firms were being forced to add service components to their product offerings. As these changes occurred and services were being established as an integral aspect of doing business, the United States had become a service economy. Ultimately, the debate over whether services were really different from physical products produced the rationale that the skeptics demanded. The evidence and essence of this argument epitomizes the service orientation.

Characteristics of all Services

There are certain fundamental characteristics that are inherent in all service products and in the service components of any product. Let us begin by defining service as *a deed, performance, or action*. Thus, by definition, services are intangible. The product that is a service or that component of the product that is a service cannot be seen, touched, or felt. As a deed, performance or action, a service is consumed as it is produced, such that the acts of production and consumption are inseparable. Thus, services are also perishable, in that they cannot be inventoried or produced and stored for later use. Nor can they be produced without some level of customer interaction.

Since services are produced and consumed in real time, they are inherently variable — from customer to customer, from provider to provider, and from time to time for the same customer and/or the same provider. These statements describe features that are fundamental and are inherent in products or components of products that are deeds, performances, or actions, and may sound very simplistic until you delve into the meanings and consequences of these inherent characteristics to the customer and to the provider.

It is generally accepted that services are different from physical goods along four fundamental characteristics labeled as intangibility, simultaneity or inseparability, perishability, and variability. These characteristics are conceptually inherent in all services or in the service component of any product. Although this framework is more useful as a pedagogical vehicle rather than a framework for research or practice, it is powerful in fully capturing the concept of service orientation and accomplishes the objective of placing the reader in the necessary frame of mind.

This mindset is grounded in the appreciation of the fundamental nature of services necessitating appropriate managerial decisions and actions. The fundamental character of services and the associated consequences may play out for the manager. A service orientation requires a thorough understanding of how these inherent characteristics of services are manifested for the customer and the manager.

Intangibility — Services

Services are performances. Services cannot be seen but they can be experienced. The product is a process. As a service provider, you cannot show your product as you could if you were the marketer of a packaged good. Yes, services are the result of value-creating activities, just as physical products are, and may employ tangibles or physical products in producing the service. However, the product being purchased is an experience and not a physical good. Customers cannot take ownership or title to a service. For example, hotels provide the service of overnight stay as their core product. The customer does not take title to the room that is being rented. The hotel provides the use of the room for the duration of the time that the customer has paid for. Similarly, customers don't take title to anything in air travel or in entertainment. When office copiers come with service, there is no ownership involved with the service component unlike with the copier itself. When your automobile comes with free service under warranty, unlike the automobile itself there is no real ownership of the service component.

Now, as a manager, you might say: "All this is well and good, but if it doesn't change the way I manage the product, its production, or its marketing, whether it is a packaged good or service, why should I care? Services are intangible and service components in products are intangible—yes! But so what?" Let us examine this question from the customer's point of view. For customers, one immediate consequence of the intangibility is increased perceived risk. When you cannot see the product or what you are going to get before the purchase, customers have to acknowledge a certain amount of risk they are taking. While there is perceived risk in the purchase of a packaged or physical good, you cannot return a vacation as you can a defective lawn mower.

Thus, you accept a certain amount of risk as unavoidable in the case of a service. When you book your vacation, do you really see the product? If you have previously been to the locations, you may have seen the facility, but your product was the experience. Complicating the perceived risk is the fact that the evaluation of services or the service component of the product is inherently subjective. Compared to packaged goods, customers find it harder to evaluate services before the purchase. In some cases, services are harder to evaluate even during and after their performances.

In the case of health care, for instance, you use several proxy elements, such as the cleanliness of the facility, the medical professional's "bedside manner, " and the process you had to go through. What you are evaluating is a lot more than the core product, the medical treatment itself. Being intangible, services cannot be produced until you have purchased (whether you pay before, during, or after) the service.

Simultaneity—Services

The acts of production and consumption occur simultaneously in services. This is primarily because services are produced and consumed in real time. There is a great deal of interaction before, during, and/or after the service between provider and customer. Some sort of customer interaction is necessary even if customer physical presence is not. At the very least, customers have to specify their needs and their need situations. In most if not all cases, it can be argued that production and consumption cannot be temporally separated. (Thus, this characteristic is also termed "inseparability.") A related complication for services is that in services, the provider is part of the product.

As a customer, the frontline personnel you interact with are a part of the product. In professional services, for instance, the lawyer is part of the (legal service) product, the doctor is part of the (medical care) product, the professor is part of the (education) product.

Comparing this to packaged goods: Do you have to interact with Procter & Gamble and the shop floor employee who made your particular tube of Crest toothpaste? You do interact with the retailer of the toothpaste, but remember that the retail store that makes the toothpaste available to you is a service. What you are consuming from the retailer as it is produced by the retailer is a service. In fact, you interact with the provider of the service component of the packaged good. If you interacted with P&G, it was with customer service.

Once again, let us confront the question: "So what?" To understand what difference it makes for the manager, the customer-focused firm must examine the consequences to the customer from the customer's perspective. The customer has to interact with the provider at some stage or at all stages of the production process. The customer is in the service factory in the case of an amusement park. Sometimes the service factory comes to the customer, as in the case of the landscape contractor. Sometimes the interaction is at arm's length, such as in the case of an online travel agent like Expedia.com. In each case, there is some interaction that the customer usually initiates. Thus, there is some effort on the part of the customer for the product to be produced.

The customer takes on a role in the production process. To perform the expected role, customers have to be educated and sometimes socialized into the process. For example, at a fast food restaurant, the customer needs to get familiar with the process of ordering and picking up the food. When you call your long distance phone company for a question on your bill, you need to have some information ready for the service to be performed. These days you have to be familiar with complex automated voice menus before you can get anything done over the telephone. Since consumption and production are simultaneous, the customer is consuming as the product is being produced. The product cannot be produced ahead of time and then consumed. This means that the product cannot be inventoried.

Perishability—Services

Since services cannot be produced and stored for later use, as physical products can, services are said to be perishable products. What actually perishes? In fact, what perishes is the productive capacity of the service, or more precisely, the opportunity to produce a product. A hairdresser's time is wasted or unproductive when he or she is not serving a customer. The customer service personnel ready to handle customer enquiries is not producing a service unless there is a customer to serve. An empty airline seat perishes without a passenger in it on takeoff. The factors of production such as labour, facilities, equipment, and billable time are the value-creating assets for the service provider.

The opportunity to produce the product from these assets perishes without the adequate number of customers for which the capacity is designed. And the service provider needs to maintain a certain level of capacity that is not easily adjustable.

Some services are more capacity-constrained than others. A hotel cannot reduce the number of rooms in its facility when there are not enough guests. A management consulting firm with fulltime employees would be hard pressed to release its staff when there are not enough clients to fill the capacity.

For the customer, service products need to be available and accessible when and where they are needed. When the productive capacity is sometimes not able to meet the demand, customers are likely to have to wait in queues or find another provider. Whether it is on the phone for customer service, at the doctor's office, or in line for a ride in an amusement park, customers will have to get used to waiting. In some services, customers need to plan ahead of time and place reservations. That is how managers instinctively manage the utilization of capacity in the operation.

Conversely, when there is more productive capacity than there are customers to be served, service providers have to look for other customer segment opportunities for their value-creating assets. Service providers have to manage that balance between capacity and demand. They employ methods to anticipate and manage the pattern of demand and to manage the capacity accordingly.

Variability — Services

Service products as experiences vary from one experience to the next, from customer to customer, as well as for the same customer from one occasion to the next. In fact, this variability is compounded with differences among frontline personnel. When there are several steps in the service process where the customer interacts with several different personnel of the service provider, there could be variation from one interaction to another. In some cases, customers will prefer specific frontline personnel with whom they have become familiar and comfortable. Since services are produced and consumed in real time, it is clear that customers will likely see a great degree of variation in product quality.

Customers can receive a different experience each time, even at a standardized operation like McDonald's. Service managers attempt to deliver consistently high quality with frontline training and technology and to customize where possible and standardize where necessary. Customers expect customization even when it is not feasible. Managers have to balance the economics of standardization with the quality issues around customization.

Thus, the fundamental characteristics of services force their implications on the customer. The manager of services and products that involve a significant degree of tangibles are forced to orient themselves to these implications. The complexity of managerial situations as a consequence of these characteristics is manifested as opportunities and challenges. Consequently, the services manager must analyze these consequences, and be prepared to take some actions to address them.

MANAGERIAL OPPORTUNITIES AND CHALLENGES

The implications of the fundamental characteristics of services suggest that managers need to examine how they would adapt their business processes to take advantage of the opportunities and to address the challenges. The following is a discussion of some of the key managerial challenges and opportunities.

Reducing Customers' Perceived Risk

With products that contain a significant proportion of intangibles and where the core product is a service, managers of services need to continuously monitor and understand the risk customers feel when they purchase the service. What is the nature of the risk? How do you lower this risk? Professional services providers such as lawyers and doctors have certifications prominently displayed on their walls. Consulting firms provide lists of clients, mutual funds companies show off their performance charts. What would be most appropriate for you to do to reduce customer perceived risk? To determine this you begin by understanding what those risk levers might be. What is the source of the anxiety that customers feel? To what would you attribute the risk that they feel?

Only when this is clearly understood can you think about the possible ways you might reduce their perceived risk. To understand the risk, you need to break down the service into the features that contribute value toward the total solution. Which service features produce benefits, and how can the risk associated with each feature-benefit connection be reduced? The degree to which the customer feels assured that expectations of these benefits will be met would depend on how you can instill confidence in the customer that you have the ability to deliver on those expectations. Christopher Hart is credited with the concept of service guarantees that works to reduce perceived risk. L.L. Bean's famed "100per cent satisfaction guaranteed" goes back to 1912, when the founder was marketing the Maine hunting shoe. Service providers who understand their customers' perceived risk seek to address it.

Customization versus Standardization

One of the powerful implications of the fundamental characteristics of services is that they are not produced ahead of time. The service is designed, created, and delivered to customer specifications. While customization is possible, some standardized parameters are also necessary to derive economies of scale. There is opportunity to standardize as well as to customize. Where there is low margin from a service component and the volume provides the profits, perhaps ways to standardize the production process would be appropriate. On the other hand, if there is opportunity for high margin, customization and low volume is the way to go.

Consider the difference between automated telephone-based customer service and a fee sometimes charged for customers who want individual attention. Professional or knowledge based services are generally customized and are high-margin, whereas dry cleaning services are low-margin services and require considerable standardization and the required volume. The opportunity is to increase margins by providing customized experiences and the challenge is in standardizing the process to reduce costs and maintain consistent quality.

Quality Consistency

A related issue to the opportunity to customize is the accompanying challenge posed by the variability of the product. Since services are produced in real time, and they have a high degree of variability, the challenge is to ensure a consistent quality in the service. Managers must assess the sources of variation. Sources of the variability are typically associated with the people delivering the service. Every interaction with the customer provides or facilitates some aspect of the product bundle, and therefore the quality of the product is dependent on each and every interaction between the service provider and the customer. Therefore, selecting, training, and motivating frontline personnel become very critical.

For example, service firms such as Walt Disney amusement parks and the Ritz-Carlton hotels have a specific profile in mind when they hire people for the front line. These firms provide extensive service training for their frontline employees, and they motivate employees to provide customer satisfaction. FedEx rewards employees who take care of customers in service failures. Empowered frontline personnel are more likely to deliver quality service.

Demand and Capacity Management

Since services are perishable, productive factors that are in place may be idle at low demand times and stretched to capacity at high demand times. The demand is fluctuating, and it might be impossible or at best a serious challenge to stretch or contract capacity. Therefore, smoothing demand and managing capacity become a constant endeavor for the managers of services. The key to managing this challenge is to understand the demand patterns and to figure out the flexibilities in your capacity. To the extent that the demand patterns can be predicted with a good understanding of customer behaviour, service providers can manage demand as well as capacity. Pricing is commonly used in smoothing demand.

Early bird specials, matinee prices, lower fares if you stay over Saturdays are all different ways by which service firms use pricing to manage demand. Reservations and appointments are another way for service managers to smooth demand. Different segments and markets are served at different times and prices to fill capacity. Hotels depend to a great extent on the lucrative conferences and meetings market to ensure capacity utilization.

Such capacity-constrained services rely more on managing demand than on managing capacity primarily because it is harder to adjust capacity. Product mix adjustments are another way of managing capacity, whereas the more versatile the value-creating assets are in producing different services with the same productive factors, the better the capacity management opportunities. For example, ski resorts turn into hiking or whitewater rafting facilities in the summer. Staffing and scheduling are typical capacity-side management options that the service firm employs. Creative use of value producing assets is critical for efficient and effective management of a firm's capacity.

Managing the Customer

Since the customer is part of the process and some degree of interaction with the customer is necessary and inevitable, managers of services must find ways to educate customers on their roles. If the customer fails in playing the appropriate coproducer role, it could ruin the process and the outcome quality. A clear understanding of the steps in the process and how, where, and when customers interact with the process is a must. Doctors, nurses, and pharmacists provide directions and instructions for patient roles. Banks and other financial institutions have specific protocols for various financial transactions.

A key question to ask is how the customer should participate in the process and whether you have the appropriate instructions for your customer. Differences in customer willingness and ability to participate in the production of the process force the issue of segmentation on the service delivery process. Thus, restaurants have self-service, take-out, or drive-through options. Automatic teller machines are an early example of self-service in banking services. Hotels have express check-in and check-out facilities. Airlines have electronic ticketing options. Self-service checkout facilities have become a standard option at grocery stores in the United States and elsewhere.

It is clear as we think about these issues that not all services are alike. While all services face all these challenges to some degree, the challenges manifest themselves differently in different services. In fact, the management solutions presented are more applicable in some and inappropriate or impossible in others. There is a systematic way of addressing these differences among services by examining characteristics that differentiate services from one another. A service that shares characteristics with another service exhibits similar challenges and opportunities.

CATEGORY CHARACTERISTICS OF A SPECIFIC SERVICE

In employing the service orientation, every manager needs to understand that there are different types of services in every product, each with its own manifestations of the fundamental characteristics of services. The nature of these manifestations in different services implies different situations for the customer and the manager. While all services are fundamentally intangible, inseparable, perishable, and variable, services differ from each other along dimensions peculiar to their category. These category characteristics provide the manager with an insight into the specific situations that they have to deal with in their particular service category.

Christopher Lovelock demonstrated that this sort of categorizing is necessary and powerful as a means of understanding the managerial situations specific to each type of service. Service managers should look toward other services that share similar characteristics and expect to find similar challenges and transferable solutions to those challenges. Consider for example that hospitals provide executive-style rooms for the businessperson who is a patient. They learned this from hotels that have executive-style suites for the businessperson. Now consider the service characteristics that are shared between hospitals and hotels.

In both cases, the customer comes to the factory and frequently requires an overnight stay. In both situations, the executive is likely to be away from an office with all its business support for an extended period of time. Both services are similar in the type of customer interactions, even though they are in two completely different service sectors. An analysis represents the benefits of learning from the successes of services that share certain specific service characteristics.

Managers of services can understand the specific challenges and opportunities presented by the characteristics of their product and corresponding situations by examining a number of characteristics that profile a service category. Here are a few characteristics to examine:

- Tangible or intangible act
- Person or thing as recipient of the act
- Type of interaction between provider and customer
- Service delivered by people or equipment
- Extent of customization or standardization

What are the tangible and intangible acts involved in the services that you are providing? Is the service act delivered on a person or a thing? For example, if you are an entertainment electronics repair and service provider, you perform tangible acts on a DVD player and intangible acts with the customer as you obtain information on what problems the customer has experienced with the DVD player. Now, examine the nature of the acts and the impact on the customer. When you are working on the DVD player, the customer needs not be present and the process that you employ need not be visible to the customer. However, you do interact with the customer at a number of points, from the initial customer call to the final interaction at delivery or pickup of the DVD player.

Similarly, you might analyze your service about the nature of interaction of with the customer. Is it face-to-face or arm's length, via telephone or the Internet? Do you visit the customer or does the customer visit your premises? How much technology is involved in the interface with the customer? Sometimes the customer does not even interact with a person. How much standardization is built into the service? Are there options provided to customers who require a customized interaction? What benefits does this analysis provide the manager?

The challenges and opportunities discussed earlier as fundamental are common to all service managers. By understanding the characteristics of their specific category of services, the manager can find ways to take advantage of the opportunities and address the challenges of their specific situation. Additionally, in understanding how the fundamental characteristic of intangibility manifests itself in their particular situation, managers can understand customers' perceived risk and use the tangibles in the customer experience to reduce the risk that customers perceive.

By understanding the nature of the interaction, managers are in a better place to manage the roles of customers and recognize opportunities for customization and standardization to improve the customer experience and, therefore, customer value. What is most interesting is that you may find that service businesses in a different industry sharing certain characteristics may offer you innovative ideas to meet these challenges and capitalize on opportunities to better serve customers in your business.

In order to understand how the required service-oriented frame of mind needs to be handled in the customer-focused firm, the firm analyzes the various services with which it augments the product. Each service is analyzed from a fundamental and category perspective. The fundamental nature of services requires an analysis of the implications of the intangible, inseparable, perishable, and variable nature of services for managerial action.

The analysis of category characteristics specific to each type of service should deal with how these consequences of these characteristics manifest themselves in that situation. The customer-focused analysis should cover the nature and the recipient of the act, the nature of the interaction, and the fluctuation of the demand.

Chapter 2

Hotel Industries and Services

NATURE OF THE INDUSTRY

Hotels and other accommodations are as diverse as the many family and business travellers they accommodate. The industry includes all types of lodging, from upscale hotels to RV parks. Motels, resorts, casino hotels, bed-and-breakfast inns, and boarding houses also are included. In fact, in 2004 nearly 62,000 establishments provided overnight accommodations to suit many different needs and budgets.

Establishments vary greatly in size and in the services they provide. *Hotels* and *motels* comprise the majority of establishments and tend to provide more services than other lodging places. There are five basic types of hotels — *commercial, resort, residential, extended-stay,* and *casino.* Most hotels and motels are *commercial* properties that cater mainly to business people, tourists, and other travellers who need accommodations for a brief stay. Commercial hotels and motels usually are located in cities or suburban areas and operate year round. Larger properties offer a variety of services for their guests, including a range of restaurant and beverage service options — from coffee bars and lunch counters to cocktail lounges and formal fine-dining restaurants.

Some properties provide a variety of retail shops on the premises, such as gift boutiques, newsstands, drug and cosmetics counters, and barber and beauty shops. An increasing number of full-service hotels now offer guests access to laundry and valet services, swimming pools, and fitness centres or health spas. A small, but growing, number of luxury hotel chains also manage condominium units in combination with their transient rooms, providing both hotel guests and condominium owners with access to the same services and amenities. Larger hotels and motels often have banquet rooms, exhibit halls, and spacious ballrooms to accommodate conventions, business meetings, wedding receptions, and other social gatherings.

Conventions and business meetings are major sources of revenue for these hotels and motels. Some commercial hotels are known as conference hotels — fully self-contained entities specifically designed for meetings. They provide physical fitness and recreational facilities for meeting attendees, in addition to state-of-the-art audiovisual and technical equipment, a business centre, and banquet services.

Resort hotels and *motels* offer luxurious surroundings with a variety of recreational facilities, such as swimming pools, golf courses, tennis courts, game rooms, and health spas, as well as planned social activities and entertainment. Resorts typically are located in vacation destinations or near natural settings, such as mountains, the seashore, theme parks, or other attractions. As a result, the business of many resorts fluctuates with the season.

Some resort hotels and motels provide additional convention and conference facilities to encourage customers to combine business with pleasure. During the off season, many of these establishments solicit conventions, sales meetings, and incentive tours to fill their otherwise empty rooms; some resorts even close for the off-season.

Residential hotels provide living quarters for permanent and semi permanent residents. They combine the comfort of apartment living with the convenience of hotel services. Many have dining rooms and restaurants that also are open to residents and to the general public.

Extended-stay hotels combine features of a resort and a residential hotel. Typically, guests use these hotels for a minimum of 5 consecutive nights. These facilities usually provide rooms with fully equipped kitchens, entertainment systems, ironing boards and irons, office space with computer and telephone lines, fitness centres, and other amenities.

Casino hotels provide lodging in hotel facilities with a casino on the premises. The casino provides table wagering games and may include other gambling activities, such as slot machines and sports betting. Casino hotels generally offer a full range of services and amenities and also may contain conference and convention facilities.

In addition to hotels and motels, *bed-and-breakfast inns, recreational vehicle (RV) parks, campgrounds,* and *rooming and boarding houses* provide lodging for overnight guests. *Bed-and-breakfast inns* provide short-term lodging in private homes or small buildings converted for this purpose and are characterized by highly personalized service and inclusion of breakfast in the room rate. Their appeal is quaintness, with unusual service and decor.

RV parks and campgrounds cater to people who enjoy recreational camping at moderate prices. Some parks and campgrounds provide service stations, general stores, shower and toilet facilities, and coin-operated laundries. While some are designed for overnight travellers only, others are for vacationers who stay longer. Some camps provide accommodations, such as cabins and fixed campsites, and other amenities, such as food services, recreational facilities and equipment, and organized recreational activities. Examples of these overnight camps include children's camps, family vacation camps, hunting and fishing camps, and outdoor adventure retreats that offer trail riding, white-water rafting, hiking, fishing, game hunting, and similar activities.

Other short-term lodging facilities in this industry include *guesthouses,* or small cottages located on the same property as a main residence, and *youth hostels* — dormitory-style hotels with few frills, occupied mainly by students traveling on limited budgets. Also included are *rooming and boarding houses*, such as fraternity houses, sorority houses, off-campus dormitories, and workers' camps. These establishments provide temporary or longer term accommodations that may serve as a principal residence for the period of occupancy. These establishments also may provide services such as housekeeping, meals, and laundry services.

In recent years, hotels, motels, camps, and recreational and RV parks affiliated with national chains have grown rapidly. To the traveller, familiar chain establishments represent dependability and quality at predictable rates. National corporations own many chains, although many properties are independently owned but affiliated with a chain through a franchise agreement.

Many independently operated hotels and inns participate in national reservations services, thereby appearing to belong to a larger enterprise. Also, many hotels join local chambers of commerce, boards of trade, convention and tourism bureaus, or regional recreation associations in order support and promote tourism in their area.

Increases in competition and in the sophistication of travellers have induced the chains to provide lodging to serve a variety of customer budgets and accommodation preferences. In general, these lodging places may be grouped into properties that offer luxury, all-suite, moderately priced, and economy accommodations. The numbers of limited-service or economy chain properties—economy lodging without extensive lobbies, restaurants, or lounges—have been growing. These properties are not as costly to build and operate. They appeal to budget-conscious family vacationers and travellers who are willing to sacrifice amenities for lower room prices.

While economy chains have become more prevalent, the movement in the hotel and lodging industry is towards more extended-stay properties. In addition to fully equipped kitchenettes and laundry services, the extended-stay market offers guest amenities such as in-room access to the Internet and grocery shopping. This segment of the hotels and other accommodations industry has eliminated traditional hotel lobbies and 24-hour front desk staffing, and housekeeping is usually done only about once a week. This helps to keep costs to a minimum.

All-suite facilities, especially popular with business travellers, offer a living room or sitting room in addition to a bedroom. These accommodations are aimed at travellers who require lodging for extended stays, families traveling with children, and business people needing to conduct small meetings without the expense of renting an additional room.

Increased competition among establishments in this industry has spurred many independently owned and operated hotels and other lodging places to join national or international reservation systems, which allow travellers to make multiple reservations for lodging, airlines, and car rentals with one telephone call. Nearly all hotel chains operate online reservation systems through the Internet.

WORKING CONDITIONS

Work in hotels and other accommodations can be demanding and hectic. Hotel staffs provide a variety of services to guests and must do so efficiently, courteously, and accurately. They must maintain a pleasant demeanor even during times of stress or when dealing with an impatient or irate guest. Alternately, work at slower times, such as the off-season or overnight periods, can seem slow and tiresome without the constant presence of hotel guests. Still, hotel workers must be ready to provide guests and visitors with gracious customer service at any hour.

Because hotels are open around the clock, employees frequently work varying shifts or variable schedules. Employees who work the late shift generally receive additional compensation. Many employees enjoy the opportunity to work part-time, nights or evenings, or other schedules that fit their availability for work and the hotel's needs.

Hotel managers and many department supervisors may work regularly assigned schedules, but they also routinely work longer hours than scheduled, especially during peak travel times or when multiple events are scheduled. Also, they may be called in to work on short notice in the event of an emergency or to cover a position. Those who are self-employed, often owner-operators, tend to work long hours and often live at the establishment.

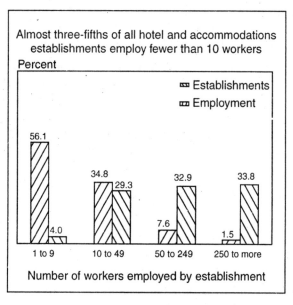

Almost three-fifths of all hotel and accommodations establishments employ fewer than 10 workers

Percent

- ▨ Establishments
- ▨ Employment

Number of workers employed by establishment			
1 to 9	10 to 49	50 to 249	250 to more
56.1 / 4.0	34.8 / 29.3	7.6 / 32.9	1.5 / 33.8

Food preparation and food service workers in hotels must withstand the strain of working during busy periods and being on their feet for many hours. Kitchen workers lift heavy pots and kettles and work near hot ovens and grills. Job hazards include slips and falls, cuts, and burns, but injuries are seldom serious. Food service workers often carry heavy trays of food, dishes, and glassware. Many of these workers work part time, including evenings, weekends, and holidays.

Office and administrative support workers generally work scheduled hours in an office setting, meeting with guests, clients, and hotel staff. Their work can become hectic processing orders and invoices, dealing with demanding guests, or servicing requests that require a quick turnaround, but job hazards typically are limited to muscle and eye strain common to working with computers and office equipment.

In 2003, work-related injuries and illnesses averaged 6.7 for every 100 full-time workers in hotels and other accommodations, compared with 5.0 for workers throughout private industry. Work hazards include burns from hot equipment, sprained muscles and wrenched backs from heavy lifting, and falls on wet floors.

Employment

Hotels and other accommodations provided 1.8 million wage and salary jobs in 2004. In addition, there were about 33,000 self-employed and unpaid family workers in the industry, who worked in bed-and-breakfast inns, camps, and small motels.

Employment is concentrated in densely populated cities and resort areas. Compared with establishments in other industries, hotels, motels, and other lodging places tend to be small. About 91 percent employed fewer than 50 people; about 56 percent employ fewer than 10 workers (chart). As a result, lodging establishments offer opportunities for those who are interested in owning and running their own business. Although establishments tend to be small, the majority of jobs are in larger hotels and motels with more than 100 employees.

Hotels and other lodging places often provide first jobs to many new entrants to the labour force. As a result, many of the industry's workers are young. In 2004, about 19 percent of the workers were younger than age 25, compared with about 14 percent across all industries.

Table: Percent distribution of employment,
by age group, 2004

Age group	Hotels and other accommodations	All industries
Total	100.0%	100.0%
16-19	5.3	4.2
20-24	13.7	9.9
25-34	22.4	21.8
35-44	23.7	24.8
45-54	20.2	23.3
55-64	11.4	12.4
65 and older	3.3	3.5

Occupations in the Industry

The vast majority of workers in this industry — more than 8 out of 10 in 2004 — were employed in service and office and administrative support occupations. Workers in these occupations usually learn their skills on the job. Postsecondary education is not required for most entry-level positions; however, college training may be helpful for advancement in some of these occupations.

For many administrative support and service occupations, personality traits and a customer-service orientation may be more important than formal schooling. Traits most important for success in the hotel and motel industry are good communication skills; the ability to get along with people in stressful situations; a neat, clean appearance; and a pleasant manner.

Service occupations, by far the largest occupational group in the industry, account for 65 percent of the industry's employment. Most service jobs are in housekeeping occupations — including maids and housekeeping cleaners, janitors and cleaners, and laundry workers — and in food preparation and service jobs — including chefs and cooks, waiters and waitresses, bartenders, fast food and counter workers, and various other kitchen and dining room workers. The industry also employs many baggage porters and bellhops, gaming services workers, and grounds maintenance workers.

Workers in *cleaning* and *housekeeping occupations* ensure that the lodging facility is clean and in good condition for the comfort and safety of guests. *Maids and housekeepers* clean lobbies, halls, guestrooms, and bathrooms. They make sure that guests not only have clean rooms, but have all the necessary furnishings and supplies. They change sheets and towels, vacuum carpets, dust furniture, empty wastebaskets, and mop bathroom floors. In larger hotels, the housekeeping staff may include assistant housekeepers, floor supervisors. housekeepers, and executive housekeepers. *Janitors* help with the cleaning of the public areas of the facility, empty trash, and perform minor maintenance work.

Table: Employment of wage and salary workers in hotels and other accommodations by occupation, 2004 and projected change, 2004-14. (Employment in thousands)

Occupation	Employment, 2004	Percent change, 2004-14	
	Number	Percent	
Total, all occupations	1,796	100.0	16.9
Management, business, and financial occupations	99	5.5	26.6

Table: Employment of wage and salary workers in hotels and other accommodations by occupation, 2004 and projected change, 2004-14. (Employment in thousands)

Top executives	16	0.9	25.8
Food service managers	10	0.6	16.2
Lodging managers	28	1.6	27.4
Meeting and convention planners	7	0.4	27.3
Service occupations	1,169	65.1	16.0
Security guards and gaming surveillance officers	34	1.9	-2.3

Chefs and head cooks	13	0.7	16.9
First-line supervisors/managers of food preparation and serving workers	22	1.2	16.5
Cooks, restaurant	56	3.1	16.7
Food preparation workers	23	1.3	27.1
Bartenders	39	2.2	13.1
Fast food and counter workers	27	1.5	25.2
Waiters and waitresses	133	7.4	9.5
Food servers, nonrestaurant	39	2.2	11.9
Dining room and cafeteria attendants and bartender helpers	43	2.4	9.1
Dishwashers	38	2.1	8.3
Hosts and hostesses, restaurant, lounge, and coffee shop	21	1.2	9.1
Supervisors, building and grounds cleaning and maintenance workers	37	2.0	26.6
Janitors and cleaners, except maids and housekeeping cleaners	49	2.7	20.2
Maids and housekeeping cleaners	405	22.5	17.0
Landscaping and groundskeeping workers	23	1.3	20.3
Gaming supervisors	11	0.6	10.3
Gaming dealers	35	2.0	25.0
Baggage porters and bellhops	25	1.4	21.5
Concierges	7	0.4	17.0
Recreation and fitness workers	13	0.7	22.1
Sales and related occupations	54	3.0	18.3
Cashiers, except gaming	16	0.9	14.3
Gaming change persons and booth cashiers	10	0.6	7.6
Office and administrative support occupations	320	17.8	15.0
Supervisors, office and administrative support workers	22	1.2	7.7
Bookkeeping, accounting, and auditing clerks	24	1.4	14.6

Table: Employment of wage and salary workers in hotels and other accommodations by occupation, 2004 and projected change, 2004-14. (Employment in thousands)

Gaming cage workers	5	0.3	5.7
Hotel, motel, and resort desk clerks	183	10.2	17.4
Reservation and transportation ticket agents and travel clerks	13	0.7	15.7
Installation, maintenance, and repair occupations	75	4.2	26.8
Maintenance and repair workers, general	64	3.6	27.2
Production occupations	39	2.2	19.0
Laundry and dry-cleaning workers	32	1.8	18.0
Transportation and material moving occupations	24	1.3	7.0

Note: May not add to totals due to omission of occupations with small employment

Workers in the various food service occupations deal with customers in the dining room or at a service counter. Waiters and waitresses take customers' orders, serve meals, and prepare checks. In restaurants, they may describe chef's specials and suggest appropriate wines. In smaller establishments, they often set tables, escort guests to their seats, accept payment, and clear tables. They also may deliver room service orders to guests. In larger restaurants, some of these tasks are assigned to other workers.

Hosts and hostesses welcome guests, show them to their tables, and give them menus. Bartenders fill beverage orders for customers seated at the bar or from waiters and waitresses who serve patrons at tables. Dining room and cafeteria attendants and bartender helpers assist waiters, waitresses, and bartenders by clearing, cleaning, and setting up tables, replenishing supplies at the bar, and keeping the serving areas stocked with linens, tableware, and other supplies. Counter attendants take orders and serve food at fast-food counters and in coffee shops; they also may operate the cash register.

Cooks and food preparation occupations prepare food in the kitchen. Beginners may advance to more skilled food preparation jobs with experience or specialized culinary training. Chefs and cooks generally prepare a wide selection of dishes, often cooking individual servings to order. Larger hotels employ cooks who specialize in the preparation of many different kinds of food. They may have titles such as salad chef, grill chef, or pastry chef.

Individual chefs may oversee the day-to-day operations of different kitchens in a hotel, such as a fine-dining full-service restaurant, a casual or counter-service establishment, or banquet operations. Chef positions generally are attained after years of experience and, sometimes, formal training, including apprenticeships. Larger establishments also employ executive chefs and food and beverage directors who plan menus, purchase food, and supervise kitchen personnel for all of the kitchens in the property. Food preparation workers shred lettuce for salads, cut up food for cooking, and perform simple cooking steps under the direction of the chef or head cook.

Many full-service hotels employ a uniformed staff to assist arriving and departing guests. Baggage porters and bellhops carry bags and escort guests to their rooms. Concierges arrange special or personal services for guests. They may take messages, arrange for babysitting, make restaurant reservations, provide directions, arrange for or give advice on entertainment and local attractions, and monitor requests for housekeeping and maintenance. Doorkeepers help guests into and out of their cars, summon taxis, and carry baggage into the hotel lobby.

Hotels also employ the largest percentage of gaming services workers because much of gaming takes place in casino hotels. Some gaming services positions are associated with oversight and direction—supervision, surveillance, and investigation—while others involve working with the games or patrons themselves, by tending the slot machines, handling money, writing and running tickets, dealing cards, and performing related duties.

Office and administrative support positions accounted for 18 percent of the jobs in hotels and other accommodations in 2004. Hotel desk clerks, secretaries, bookkeeping and accounting clerks, and telephone operators ensure that the front office operates smoothly. The majority of these workers are hotel, motel, and resort desk clerks. They process reservations and guests' registration and checkout, monitor arrivals and departures, handle complaints, and receive and forward mail.

The duties of hotel desk clerks depend on the size of the facility. In smaller lodging places, one clerk or a manager may do everything. In larger hotels, a larger staff divides the duties among several types of clerks. Although hotel desk clerks sometimes are hired from the outside, openings usually are filled by promoting other hotel employees such as bellhops and porters, credit clerks, and other administrative support workers.

Hotels and other lodging places employ many different types of *managers* to direct and coordinate the activities of the front office, kitchen, dining room, and other departments, such as housekeeping, accounting, personnel, purchasing, publicity, sales, security and maintenance. Managers make decisions on room rates, establish credit policy, and have ultimate responsibility for resolving problems. In smaller establishments, the manager also may perform many of the front-office clerical tasks. In the smallest establishments, the owners — sometimes a family team — do all the work necessary to operate the business.

Lodging managers or *general and operations managers* in large hotels often have several assistant managers, each responsible for a phase of operations. For example, *food and beverage managers* oversee restaurants, lounges, and catering or banquet operations. *Rooms managers* look after reservations and occupancy levels to ensure proper room assignments and authorize discounts, special rates, or promotions. Large hotels, especially those with conference centres, use an executive committee structure to improve departmental communications and coordinate activities. Other managers who may serve on a hotel's executive committee include *public relations* or *sales managers, human resources directors, executive housekeepers,* and *heads of hotel security.*

Workers at vacation and recreational camps may include camp counselors who lead and instruct children and teenagers in outdoor-oriented forms of recreation, such as swimming, hiking, horseback riding, and camping. In addition, counselors at vacation and resident camps also provide guidance and supervise daily living and general socialization. Other types of campgrounds may employ trail guides for activities such as hiking, hunting, and fishing.

Hotels and other lodging places employ a variety of workers found in many other industries. Maintenance workers, such as stationary engineers, plumbers, and painters, fix leaky faucets, do some painting and carpentry, see that heating and air-conditioning equipment works properly, mow lawns, and exterminate pests. The industry also employs cashiers, accountants, personnel workers, entertainers, and recreation workers. Also, many additional workers inside a hotel may work for other companies under contract to the hotel or may provide personal or retail services directly to hotel guests from space rented by the hotel. This group includes guards and security officers, barbers, cosmetologists, fitness trainers and aerobics instructors, valets, gardeners, and parking attendants.

Training and Advancement

Although the skills and experience needed by workers in this industry depend on the specific occupation, most entry-level jobs require little or no previous training. Basic tasks usually can be learned in a short time. Almost all workers in the hotel and other accommodations industry undergo on-the-job training, which usually is provided under the supervision of an experienced employee or manager. Some large chain operations have formal training sessions for new employees; many also provide video or on-line training.

Hotel operations are becoming increasingly diverse and complex, but all positions require employees to maintain a customer-service orientation. Hoteliers recognize the importance of personal service and attention to guests; so they look for persons with positive personality traits and good communication skills when filling many guest services positions, such as desk clerk and host and hostess positions. Many hotel managers place a greater emphasis on customer service skills while providing specialized training in important skill areas, such as computer technology and software.

Vocational courses and apprenticeship programmes in food preparation, catering, and hotel and restaurant management, offered through restaurant associations and trade unions, provide training opportunities. Programmes range in length from a few months to several years. About 800 community and junior colleges offer 2-year degree programmes in hotel and restaurant management. The U.S. Armed Forces also offer experience and training in food service.

Traditionally, many hotels fill first-level manager positions by promoting administrative support and service workers — particularly those with good communication skills, a solid educational background, tact, loyalty, and a capacity to endure hard work and long hours.

People with these qualities still advance to manager jobs but, more recently, lodging chains have primarily been hiring persons with four-year college degrees in the liberal arts or other fields and starting them in trainee or junior management positions. Bachelor's and master's degree programmes in hotel, restaurant, and hospitality management provide the strongest background for a career as a hotel manager, with nearly 150 colleges and universities offering such programmes. Graduates of these programmes are highly sought by employers in this industry. New graduates often go through on-the-job training programmes before being given much responsibility. Eventually, they may advance to a top management position in a hotel, a corporate management opportunity in a large chain operation, or an investment or financial analysis position in the financial services sector.

Upper management positions, such as general manager, lodging manager, food service manager, or sales manager, generally require considerable formal training and job experience. Some department managers, such as comptrollers, purchasing managers, executive housekeepers, and executive chefs, generally require some specialized training and extensive on-the-job experience. To advance to positions with more responsibilities, managers frequently change employers or relocate within a chain to a property in another area.

For office and administrative support and service workers, advancement opportunities in the hotel industry vary widely. Some workers, such as housekeepers and janitors, generally have few opportunities for advancement. In large properties, however, some janitors may advance to supervisory positions. Hotel desk clerks, hospitality workers, and chefs sometimes advance to managerial positions. Promotional opportunities from the front office often are greater than those from any other department, because this vantage point provides an excellent opportunity to learn the establishment's overall operation. Front-office jobs are excellent entry-level jobs and can serve as a steppingstone to jobs in hospitality, public relations, advertising, sales, and management.

Advancement opportunities for chefs and cooks are better than those for most other service occupations. Cooks often advance to chef or to supervisory and management positions, such as executive chef, restaurant manager, or food service manager. Some transfer to jobs in clubs, go into business for themselves, or become instructors of culinary arts.

Nature of the Work

A comfortable room, good food, and a helpful staff can make being away from home an enjoyable experience for both vacationing families and business travellers. While most lodging managers work in traditional hotels and motels, some work in other lodging establishments, such as camps, inns, boardinghouses, dude ranches, and recreational resorts. In full-service hotels, lodging managers help their guests have a pleasant stay by providing many of the comforts of home, including cable television, fitness equipment, and voice mail, as well as specialized services such as health spas. For business travellers, lodging managers often schedule available meeting rooms and electronic equipment, including slide projectors and fax machines.

Lodging managers are responsible for keeping their establishments efficient and profitable. In a small establishment with a limited staff, the manager may oversee all aspects of operations. However, large hotels may employ hundreds of workers, and the general manager usually is aided by a number of assistant managers assigned to the various departments of the operation. In hotels of every size, managerial duties vary significantly by job title.

General managers have overall responsibility for the operation of the hotel. Within guidelines established by the owners of the hotel or executives of the hotel chain, the general manager sets room rates, allocates funds to departments, approves expenditures, and ensures expected standards for guest service, decor, housekeeping, food quality, and banquet operations. Managers who work for chains also may organize and staff a newly built hotel, refurbish an older hotel, or reorganize a hotel or motel that is not operating successfully. In order to fill entry-level service and clerical jobs in hotels, some managers attend career fairs.

Resident or hotel managers are responsible for the day-to-day operations of the property. In larger properties, more than one of these managers may assist the general manager, frequently dividing responsibilities between the food and beverage operations and the rooms or lodging services. At least one manager, either the general manager or a hotel manager, is on call 24 hours a day to resolve problems or emergencies.

Assistant managers help run the day-to-day operations of the hotel. In large hotels, they may be responsible for activities such as personnel, accounting, office administration, marketing and sales, purchasing, security, maintenance, and pool, spa, or recreational facilities. In smaller hotels, these duties may be combined into one position. Assistant managers may adjust charges on a hotel guest's bill when a manager is unavailable.

An Executive Committee made up of a hotel's senior managers advises the general manager, assists in setting hotel policy, coordinates services that cross departmental boundaries, and collaborates on efforts to ensure consistent and efficient guest services throughout the hotel. The Committee may be comprised of the department heads for housekeeping, front office, food and beverage, security, sales and public relations, meetings and conventions, engineering and building maintenance, and human resources. Executive committee members bring a different perspective of guest service to the total management objective reflecting the unique expertise and training of their positions.

Executive housekeepers ensure that guest rooms, meeting and banquet rooms, and public areas are clean, orderly, and well maintained. They also train, schedule, and supervise the work of housekeepers, inspect rooms, and order cleaning supplies.

Front office managers coordinate reservations and room assignments, as well as train and direct the hotel's front desk staff. They ensure that guests are treated courteously, complaints and problems are resolved, and requests for special services are carried out. Front office managers may adjust charges posted on a customer's bill.

Convention services managers coordinate the activities of various departments in larger hotels to accommodate meetings, conventions, and special events. They meet with representatives of groups or organizations to plan the number of rooms to reserve, the desired configuration of the meeting space, and banquet services. During the meeting or event, they resolve unexpected problems and monitor activities to ensure that hotel operations conform to the expectations of the group.

Food and beverage managers oversee all food service operations maintained by the hotel. They coordinate menus with the Executive Chef for the hotel's restaurants, lounges, and room service operations. They supervise the ordering of food and supplies, direct service and maintenance contracts within the kitchens and dining areas, and manage food service budgets.

Catering managers arrange for food service in a hotel's meeting and convention rooms. They coordinate menus and costs for banquets, parties, and events with meeting and convention planners or individual clients. They coordinate staffing needs and arrange schedules with kitchen personnel to ensure appropriate food service.

Sales or marketing directors and public relations directors oversee the advertising and promotion of hotel operations and functions, including lodging and dining specials and special events, such as holiday or seasonal specials. They direct the efforts of their staff to purchase advertising and market their property to organizations or groups seeking a venue for conferences, conventions, business meetings, trade shows, and special events. They also coordinate media relations and answer questions from the press.

Human resources directors manage the personnel functions of a hotel, ensuring that all accounting, payroll, and employee relations matters are handled in compliance with hotel policy and applicable laws. They also oversee hiring practices and standards and ensure that training and promotion programmes reflect appropriate employee development guidelines.

Finance (or revenue) directors monitor room sales and reservations. In addition to overseeing accounting and cash-flow matters at the hotel, they also project occupancy levels, decide which rooms to discount and when to offer rate specials.

Computers are used extensively by lodging managers and their assistants to keep track of guests' bills, reservations, room assignments, meetings, and special events. In addition, computers are used to order food, beverages, and supplies, as well as to prepare reports for hotel owners and top-level managers. Managers work with computer specialists to ensure that the hotel's computer system functions properly. Should the hotel's computer system fail, managers must continue to meet the needs of hotel guests and staff.

Because hotels are open around the clock, night and weekend work is common. Many lodging managers work more than 40 hours per week, and may be called back to work at any time. Some managers of resort properties or other hotels where much of the business is seasonal have other duties on the property during the off-season or find work at other hotels or in other areas.

Lodging managers experience the pressures of coordinating a wide range of activities. At larger hotels, they also carry the burden of managing a large staff and finding a way to satisfy guest needs while maintaining positive attitudes and employee morale. Conventions and large groups of tourists may present unusual problems or require extended work hours. Moreover, dealing with irate guests can be stressful. The job can be particularly hectic for front office managers during check-in and check-out times. Computer failures can further complicate processing and add to frustration levels.

Hotels increasingly emphasize specialized training. Postsecondary training in hotel, restaurant, or hospitality management is preferred for most hotel management positions; however, a college liberal arts degree may be sufficient when coupled with related hotel experience or business education. Internships or part-time or summer work experience in a hotel are an asset to students seeking a career in hotel management. The experience gained and the contacts made with employers can greatly benefit students after graduation. Most degree programmes include work-study opportunities.

Community colleges, junior colleges, and many universities offer certificate or degree programmes in hotel, restaurant, or hospitality management leading to an associate, bachelor, or graduate degree. Technical institutes, vocational and trade schools, and other academic institutions also offer courses leading to formal recognition in hospitality management. In total, more than 800 educational facilities provide academic training for would-be lodging managers. Hotel management programmes include instruction in hotel administration, accounting, economics, marketing, housekeeping, food service management and catering, and hotel maintenance engineering. Computer training also is an integral part of hotel management training, due to the widespread use of computers in reservations, billing, and housekeeping management.

More than 450 high schools in 45 States offer the Lodging Management Programme created by the Educational Institute of the American Hotel and Lodging Association. This two-year programme offered to high school juniors and seniors teaches management principles and leads to a professional certification called the "Certified Rooms Division Specialist." Many colleges and universities grant participants credit towards a post-secondary degree in hotel management.

Lodging managers must be able to get along with many different types of people, even in stressful situations. They must be able to solve problems and concentrate on details. Initiative, self-discipline, effective communication skills, and the ability to organize and direct the work of others also are essential for managers at all levels.

Persons wishing to make a career in the hospitality industry may be promoted into a management trainee position sponsored by the hotel or a hotel chain's corporate parent. Typically, trainees work as assistant managers and may rotate assignments among the hotel's departments—front office, housekeeping, or food and beverage—to gain a wide range of experiences. Relocation to another property may be required to help round out the experience and to help grow a trainee into the position.

Work experience in the hospitality industry at any level or in any segment, including summer jobs or part-time work in a hotel or restaurant, is good background for entering hotel management. Most employers require a bachelor's degree with some education in business and computer literacy, while some prefer a master's degree for hotel management positions. However, employees who demonstrate leadership potential and possess sufficient length or breadth of experience may be invited to participate in a management training programme and advance to hotel management positions without the education beyond high school.

Large hotel and motel chains may offer better opportunities for advancement than small, independently owned establishments, but relocation every several years often is necessary for advancement. The large chains have more extensive career ladder programmes and offer managers the opportunity to transfer to another hotel or motel in the chain or to the central office. Career advancement can be accelerated by the completion of certification programmes offered by various associations. These programmes usually require a combination of course work, examinations, and experience. For example, outstanding lodging managers may advance to higher level manager positions.

Lodging managers held about 58,000 jobs in 2004. Self-employed managers — primarily owners of small hotels, motels, and inns — held about 45 percent of these jobs. Companies that manage hotels and motels under contract employed many managers.

Employment of lodging managers is expected to grow about as fast as the average for all occupations through 2014. Additional job openings are expected to occur as experienced managers transfer to other occupations or leave the labour force, in part because of the long hours and stressful working conditions. Job opportunities are expected to be best for persons with college degrees in hotel or hospitality management.

Renewed business travel and domestic and foreign tourism will drive employment growth of lodging managers in full-service hotels. The numbers of economy-class rooms and extended-stay hotels also are expected to increase to accommodate leisure travellers and bargain-conscious guests.

An increasing range of lodging accommodations is available to travellers, from economy hotels which offer clean, comfortable rooms and front desk services without costly extras such as restaurants and room service, to luxury and boutique inns that offer sumptuous furnishings and personal services.

The accommodation industry is expected to continue to consolidate as lodging chains acquire independently owned establishments or undertake their operation on a contract basis. The increasing number of extended-stay hotels will moderate growth of manager jobs because these properties usually have fewer departments and require fewer managers. Also, these establishments often do not require a manager to be available 24 hours a day, instead assigning front desk clerks on duty at night some of the responsibilities previously reserved for managers.

Additional demand for managers is expected in suite hotels, because some guests — especially business customers — are willing to pay higher prices for rooms with kitchens and suites that provide the space needed to conduct small meetings. In addition, large full-service hotels — offering restaurants, fitness centres, large meeting rooms, and play areas for children, among other amenities — will continue to provide many trainee and managerial opportunities.

HOTEL, MOTEL, AND RESORT DESK CLERKS

Hotel, motel, and resort desk clerks perform a variety of services for guests of hotels, motels, and other lodging establishments. Regardless of the type of accommodation, most desk clerks have similar responsibilities. They register arriving guests, assign rooms, and check out guests at the end of their stay. They also keep records of room assignments and other registration-related information on computers. When guests check out, desk clerks prepare and explain the charges, as well as process payments.

Front-desk clerks always are in the public eye and typically are the first line of customer service for a lodging property. Their attitude and behaviour greatly influence the public's impressions of the establishment. And as such, they always must be courteous and helpful. Desk clerks answer questions about services, checkout times, the local community, or other matters of public interest. Clerks also report problems with guest rooms or public facilities to members of the housekeeping or maintenance staff for them to correct the problems. In larger hotels or in larger cities, desk clerks may refer queries about area attractions to a concierge and may direct more complicated questions to the appropriate manager.

In some smaller hotels and motels, where smaller staffs are employed, clerks may take on a variety of additional responsibilities, such as bringing fresh linens to rooms, which usually are performed by employees in other departments of larger lodging establishments. In the smaller places, desk clerks often are responsible for all front-office operations, information, and services. For example, they may perform the work of a bookkeeper, advance reservation agent, cashier, laundry attendant, and telephone switchboard operator.

Hotels are open around the clock creating the need for night and weekend work. Extended hours of operation also afford the many part-time job seekers an opportunity to find work in these establishments, especially on evenings and late-night shifts or on weekends and holidays. About half of all desk clerks work a 35 to 40 hour week—most of the rest work fewer hours—so the jobs are attractive to persons seeking part-time work or jobs with flexible schedules. Most clerks work in areas that are clean, well lit, and relatively quiet, although lobbies can become crowded and noisy when busy. Many hotels have stringent dress guidelines for desk clerks.

Desk clerks may experience particularly hectic times during check-in and check-out times or incur the pressures encountered when dealing with convention guests or large groups of tourists at one time. Moreover, dealing with irate guests can be stressful. Computer failures can further complicate an already busy time and add to stress levels. Hotel desk clerks may be on their feet most of the time and may occasionally be asked to lift heavy guest luggage.

Hotel, motel, and resort desk clerks deal directly with the public, so a professional appearance and a pleasant personality are important. A clear speaking voice and fluency in English also are essential, because these employees talk directly with hotel guests and the public and frequently use the telephone or public-address systems. Good spelling and computer literacy are needed, because most of the work involves use of a computer. In addition, speaking a foreign language fluently is increasingly helpful, because of the growing international clientele of many properties.

Most hotel, motel, and resort desk clerks receive orientation and training on the job. Orientation may include an explanation of the job duties and information about the establishment, such as the arrangement of sleeping rooms, availability of additional services, such as a business or fitness centre, and location of guest facilities, such as ice and vending machines, restaurants and other nearby retail stores. New employees learn job tasks through on-the-job training under the guidance of a supervisor or an experienced desk clerk.

They often receive additional training on interpersonal or customer service skills and on how to use the computerized reservation, room assignment, and billing systems and equipment. Desk clerks typically continue to receive instruction on new procedures and on company policies after their initial training ends.

Formal academic training generally is not required so many students take jobs as desk clerks on evening or weekend shifts or during school vacation periods. Most employers look for people who are friendly and customer-service oriented, well groomed, and display the maturity and self confidence to demonstrate good judgment. Desk clerks, especially in high-volume and higher-end properties should be quick-thinking, show initiative, and be able to work as a member of a team. Hotel managers typically look for these personal characteristics when hiring first-time desk clerks, because it is easier to teach company policy and computer skills than personality traits.

Large hotel and motel chains may offer better opportunities for advancement than small, independently owned establishments. The large chains have more extensive career ladder programmes and may offer desk clerks an opportunity to participate in a management training programme. Also, the Educational Institute of the American Hotel and Motel Association offers home-study or group-study courses in lodging management, which may help some obtain promotions more rapidly.

Hotel, motel, and resort desk clerks held about 195,000 jobs in 2004. Virtually all were in hotels, motels, and other establishments in the accommodation industry. Few were self employed.

Employment of hotel, motel, and resort desk clerks is expected to grow about as fast as the average for all occupations through 2014, as more hotels, motels, and other lodging establishments are built and occupancy rates rise. Job opportunities for hotel and motel desk clerks also will result from a need to replace workers, because many of these clerks either transfer to other occupations that offer better pay and advancement opportunities or simply leave the workforce altogether. Opportunities for part-time work should continue to be plentiful, because these businesses typically are staffed 24 hours a day, 7 days a week.

Employment of hotel and motel desk clerks should benefit from an increase in business and leisure travel. Shifts in preferences away from long vacations and toward long weekends and other, more frequent, shorter trips also should boost demand for these workers, because such stays increase the number of nights spent in hotels. While many lower budget and extended-stay establishments are being built to cater to families and the leisure traveller, many new luxury and resort accommodations also are opening to serve the upscale client. With the increased number of units requiring staff, employment opportunities for desk clerks should be good.

Growth of hotel, motel, and resort desk clerk jobs will be moderated by technology. Automated check-in and check-out procedures reduce the backlog of guests waiting for desk service and may reduce peak front desk staffing needs in many establishments. Nevertheless, the front desk remains the principal point of contact for guests at most properties and most will continue to have clerks on duty.

Employment of desk clerks is sensitive to cyclical swings in the economy. During recessions, vacation and business travel declines, and hotels and motels need fewer desk clerks. Similarly, employment is affected by special events, business and convention business, and seasonal fluctuations.

The restaurant and hotel industries consist of establishments that are open to the public or are operated by membership organizations that furnish meals or lodging. The restaurant industry is composed of establishments that prepare and serve meals and beverages and includes, but is not limited to, restaurants, cafeterias, caterers, cocktail lounges, diners, fast food places, and takeout or delivery businesses.

Establishments in the hotel industry provide lodging to their customers or members and include, but are not limited to, hotels, motels, hostels, inns, rooming and boarding houses, fraternity or sorority residential houses, and residential clubs.

An Employee

A worker is a common law employee when the employer has the right to direct and control the manner and means of accomplishing the work. Types of employees that are typical in the restaurant and hotel industries are:

- Chefs
- Dishwashers
- Bus Persons
- Waiters and Waitresses
- Hosts and Hostesses
- Managers
- Bartenders

- Cooks
- Kitchen Helpers

- Maitre d's
- Cashiers
- Delivery Persons
- Valets

- Clerical and Office Staff
- Switchboard Operators
- Repair and Maintenance Persons
- Bellhops

- Maids
- Laundry Persons
- Desk Clerks

Other services that may be performed by an employee under common law rules include, but are not limited to, those of bookkeepers, janitors, and entertainers.

Wages

Wages are payments made to an employee for services performed during employment. The payment may be made in cash or some medium other than cash. Types of payments typically considered to be wages are:

- Cash
- Meals and Beverages

- Lodging
- Tips

Employer-provided meals and lodging are subject to Unemployment Insurance (UI), State Disability Insurance (SDI), and Employment Training Tax (ETT). Meals are subject to personal income tax (PIT) withholding and reportable as PIT wages unless furnished for the employer's convenience and on the employer's premises.

If more than half of the employees receive meals that are for the convenience of the employer, all meals furnished by the employer are considered furnished for the employer's convenience and are therefore not subject to PIT withholding or reportable as PIT wages. If fewer than half of the employees receive meals which are for the convenience of the employer, only those meals actually provided for the employer's convenience would be exempt from the PIT withholding and wage reporting requirements.

Lodging is also subject to PIT unless furnished on the employer's premises, for the employer's convenience, and as a condition of employment.

The Values of Meals and Lodging

The taxable values of meals and lodging should not be less than the reasonable estimated values stipulated by the contract of employment or in a union agreement. If the cash values are not stipulated in the hiring or union agreement, the taxable values are established by regulation. The taxable value of lodging is 66 2/3 percent of the ordinary rental value to the public up to a maximum per month and not less than a minimum value per week.

The taxable values of meals and lodging are listed below:

The cash values of meals and lodging are subject to change each calendar year. This information is published in the Employment Development Department's (EDD) quarterly newsletter.

Employees who receive more than $20 in tips in a calendar month must report all tips in one or more written statements to the employer on or before the tenth day of the month following the month in which they are received from the customers. Tips are taxable when the employee's statement is furnished to the employer.

Banquet tips and tips controlled by the employer are treated as regular wages, and their taxability is not contingent upon employees reporting them to the employer. Tips that are included in a written statement furnished to the employer are wages and are subject to UI, ETT, SDI, and PIT. Tips should be combined with regular wages when reported to EDD.

A Survey Report

Using specially developed models of service management, the researchers looked at five key service sectors: retail, hotels and catering, health care, utilities, and professional and financial services. "We identified the top 10 companies that have set the standard for service quality in America over the past decade and uncovered the secrets of the global leaders,". "Each excelled by systematically linking the drivers of service excellence: leadership, people, processes and performance management. We discovered that delivering service quality goes beyond simplistic prescriptions about people issues, and instead extends to strategic factors, such as organizational design, leadership and market acuity, to orchestrate the entire service encounter."

The researchers found vast differences among the service sectors. The retail and hotel industries were the two highest performing sectors in the survey; the researchers found they follow best practices, achieving good results. The hotel industry also paid more attention to customer service than any other sector, which paid off in hefty returns. Not all hotels are wonderful, but those that succeed manage the "evidence" of quality, such as how the hotel looks and how employees dress. These hotels also manage customer interactions and continuously communicate their high standards to employees, according to the study.

Other pacesetter sectors in the study are industrial service and telecommunications companies. The researchers found that financial-services companies satisfy customers, but they concluded that the institutions may be yielding good performance without being brilliant strategists. Many are riding a wave of low interest rates and a favourable economy.

For example, some financial institutions seemed complacent about managing customer interactions, compared to their counterparts in other service sectors. By stressing re-structuring and reducing their staffs at the expense of customer service, businesses like banks and insurance companies may be vulnerable to future market changes.

Professional services such as legal and architectural firms were found to be "surprisingly weak" in both practice and performance. Of the top 10 service practices emphasized by senior executives, half are customer-focused: accessibility, listening to customers, competitive positioning, consistently meeting customers' needs and customer orientation.

Accessibility is the new "location, location, location" of service. It refers to a customer's ability to contact someone in a company easily, and at many companies, 24 hours a day. The top 10 companies were described as having "global-service leadership," demonstrating world-class capabilities, regardless of the geographic market of a company's location. Nordstrom, the retail chain, was cited as a classic example of a company that maintained world-class standards in both practice (such as how they control quality and handle complaints) and service performance (such as value, quality, satisfaction, business performance and customer retention).

The study also found that managing change is one of the biggest hurdles to achieving global service leadership. However, by world standards, the top American service businesses enjoy a supremacy role similar to that of Japan's manufacturing sector. The researchers were involved an in-depth, year-long survey of medium- to large-size service companies in the United States.

They conducted three-hour interviews with senior executives who, besides responding to 80 questions about their own company, rated other American companies on their quality of service. Participants were not allowed to identify themselves in the "best company" category. A hotel, resort, casino, or other hospitality asset's positioning is uniquely complex among real estate property types.

Changing market factors, including outside economic forces, shifts in demand patterns, and changing competitor pressures can all influence a property's success and require an owner to constantly re-evaluate an asset's current positioning. A sound repositioning strategy is an effective way to maximize a property's future financial returns.

The results show that 39 percent of the 181 companies surveyed had the potential to become global leaders in delivering good customer service, although only 13 percent of these companies have achieved both world-class performance and world-class service management practices. The research team also surveyed 21 governmental and non-profit organizations.

The results clearly show a need for improvement in governmental agencies. With the exception of altruism and cost performance, the study found that productivity, quality service and customer growth and retention lagged behind private enterprises. The researchers found most government organizations do not measure value or are not striving sufficiently to create it. Major changes in management and the approach to customer are needed to produce better results.

Hospitality & Leisure practice is comprised of industry professionals with significant experience in successful product positioning. Knowledge of the local market, the latest in competitive product offerings, and effective branding strategies are vital to creating a successful repositioning strategy. The specialists possess the knowledge and skills to formulate a creative vision for a market-based, property-specific repositioning, and employ proven methods to analyze and quantify potential financial benefits to the owner.

The professionals will not simply recommend a plan to reposition a property, but take the next step to assist the owner in understanding how the repositioning strategy should be implemented so that optimal financial returns can be achieved. The specialists investigate a multitude of possibilities for a property's potential repositioning, based on historical and recently changing market conditions, and the property's past and current positioning in the market.

Conducting a thorough market analysis and obtaining a solid understanding of the local economy, area demographics, demand generators, market trends, and other related area-specific characteristics are only the first steps taken in formulating an effective repositioning strategy. Considers every aspect of a property's operations in relation to its local market competition and the needs of the local market, to determine a relevant solution with an optimal financial payback.

Chapter 3

Managing Customer Relationships

SELECTING AND ATTRACTING THE RIGHT CUSTOMERS

The task of creating and delivering superior customer value must be complemented with the selection of the appropriate customers and the effective management of relationships with those customers. The hospitality industry is a fascinating one from a CRM perspective, because of the quality and quantity of customer touchpoints. In the world of servicing guests, there are as many challenges as there are, well... challenging customers, but in the current age of "branding" one of the biggest ones is ensuring a consistent customer service experience. Today's hotels offer a multitude of lucrative services, all of which need to be recorded immediately against the right customer's bill.

Zonal can integrate any third party billing program to ensure that all room-charged items can be cross-checked quickly and easily at the point-of-sale for matching name against a stated room number. Commonly, hotels have a number of different sales areas for patrons to drink, eat and enjoy services.

Hotel management has the ability to define different products and prices for each of these sales areas, allowing them to apply premium price bands for exclusive areas. This challenge is three-pronged: First, managers must be able to manage consistency in the face of interchanging slow and busy times and seasons. Second, consistency needs to be ensured across job titles, roles, and pay ranges.

Third (or perhaps First, if you'd like), the marketing message must be in tune with a plan to set guest expectations according to the season and customer tier considerations. Customer retention leading to more custom and bigger profit is easier if you can keep your customers happy. This can be achieved in your restaurants with minimal wait times for food orders and accurate service delivery - every time.

The fully integrated kitchen management system communicates all food orders, with special instructions if necessary, automatically from the PoS in your sales areas directly to the kitchen to minimise customer wait times. Real-time awareness of current stock holding and usage gives you the power to manage supply and monitor profit margins more effectively. The stock control system is one of the most powerful in the industry and gives you the flexibility to monitor movements of products of all divisions between the different zones in your site. Emergency transfers of champagne cases from the cocktail bar to the function room can now be recorded easily.

As with any strategy, the goal is to help meet the corporate objectives, which often begin with defining the customer segments that can help move the enterprise in the right direction, and then approach them with the right marketing message, via the right marketing channels. On the subject of job roles, while some job roles had specific training on interacting with customers, others did not, or worse, were trained in an inconsistent way.

So the first order of things in this area is to establish a clear procedure for greeting, servicing, and addressing guest issues across various situations. All employees which come in direct contact with guests need to be in tune with this common standard. A set of behaviour and service standards also provides clear guidelines which can empower employees to provide special, or "magical" moments to their guests. The guest servicing standards themselves should focus on not just consistent responses, but also should prevent consecutive negative experiences.

In other words, if a guest has experienced a negative event (i.e. complained that the room wasn't clean upon check in), this fact needs to be captured and made available to customer-facing employees so that they can put an extra effort into making sure that the remainder of the hotel stay or restaurant experience is as positive as possible. A further back-end benefit to capturing the negative event information is that it will enable analysis of customer experience shortcomings. This in turn allows for active methods for managing, monitoring, and predicting customer satisfaction, which can then lead to fine-tuning the marketing message, interaction standards, and employee training.

Predictive analytics can be used to understand latent pain or dissatisfaction before it percolates and impacts customer loyalty.Managing customer relationships should be guided by an understanding of what the customer's equity is to the firm. Customer equity, the value of customer to the firm, improves as superiority in customer value improves. Since customer equity is perhaps the firm's most valuable asset, the firm must continually seek to improve customer value for it best customers. A firm's potential return on customer equity should determine the investment it makes in customers. For its most valuable customers, the firm must guarantee superior service quality and customer service, with special attention to recovering from inevitable product failures when they occur.

Maximizing long-term profitability comes from maximizing customer equity — firms must maximize the lifetime value of the customer, including revenues, referrals as well as costs of serving the customer. Customer acquisition and retention efforts must be guided by the worth of the customer to the firm.

Maytag provides premium service to its premium customers—those who purchase the Neptune line of laundry machines. Neptune customers get a dedicated staff, a separate toll-free number and fast response on service calls. This is an example of the common business practice where firms allocate resources by the profitability and value of the customer. They utilize the opportunity in directly interacting with individual customers to determine customer profitability and allocate assets accordingly. And what are the benefits of the practice of differentiating among your customers?

Broadly referring to the practice as CRM can sometimes defeat its purpose by losing sight of the basic meaning of the term: managing customer relationships. Managing customer relationships would mean actively planning, organizing, directing, and controlling a firm's business relationships with its customers. The term might be "new" in its current usage, but the business practice of managing relationships with customers is certainly not new.

What has prompted the increased attention to CRM is new technology: how well firms can practice CRM has been advanced by information technology in the new economy. Technology brings with it the risk of missing the benefits of huge investment costs if used inappropriately, however. The benefits can seem so attractive that the costs are rationalized until the technology fails to deliver.

The American Customer Satisfaction Index, a measure of customer attitudes toward about 200 companies from over 30 different industries, has actually shown a decline, while at the same time CRM technology investments had grown about five times. When used with a fundamental understanding of CRM's purpose, the benefits of CRM technology enabling the business practice of managing customer relationships are clearly powerful. Siebel, Peoplesoft, Oracle, and other CRM technologies are really customer information management or customer knowledge management systems. These systems gather data and convert it to knowledge that will help firms in their customer relationship management activities.

The most significant contribution of CRM technology to the practice of business is not the technology itself, of course. It is in what the technology does to the practice of managing customer relationships. Because technology has now made it easy to do all the tasks of gathering and analyzing customer information, firms have been able to discover and realize the incredible benefits in proactively managing customer relationships. For example, Continental Airlines' customer information system allows its staff to mine data on passenger profitability and is also able to suggest remedies and perks for special requests or complaining customers.

Customer benefits from relationships with firms have been conceptualized under three categories: social, psychological, and customization benefits, in a two-part study using interviews and surveys. Financial services such as brokerage and banking are heavy users of CRM technology. Deregulation and information technology have effectively blurred the boundaries between those two once-different financial institutions.

They have had a heavy reliance on information because these are primarily knowledge businesses. "Signature"-level customers at Charles Schwab wait no longer than fifteen seconds to reach a customer service person, whereas other customers can wait ten minutes or more. Some banks have coded their customers so that customer service reps can decide on rates and fees depending on the customer's profitability code. Centura Banks rates its customers on a profitability scale of 1 to 5.

The most profitable customers get service calls from staff and an annual call from the CEO. Attrition rate at the bank is down 50 percent in four years, and—more interesting—the percentage of unprofitable customers has gone down from 27 percent to 21 percent. The hospitality industry was able to slash 50 percent of its promotion programmes and increase response rates by 20 percent with a good database of response behaviour from its mailing list. CRM systems have provided firms with the data they need to determine the revenues and costs to serve at the individual customer level, allowing firms to prioritize their allocations of value-creating assets and resources to the more profitable customers.

According to AMR research, the CRM market grew from $200 milllion to $1.1 billion between 1994 and 1997, and is expected to reach as much as $16 billion — an indication of how much firms want this technology to manage customer relationships. A firm's knowledge about its customers has allowed it to adjust customer value based on the profitability of the individual customer.

Just as with power, information technology has to be used judiciously. Discriminating against less profitable customers can seem unreasonable to all paying customers and could backfire with publicity. AT&T withdrew its minimum usage charges for its basic-plan customers who were unprofitable. GE Capital tried to charge credit-card users who were not accruing a minimum level of interest charges and ended up having to sell its credit card business. When used appropriately at the individual level, information technology can be very rewarding. Capital One's senior vice-president for domestic card operations, Marge Connelly, says, "We look at every single customer contact as an opportunity to make an unprofitable customer more profitable."

CRM systems are not just about profiling customers and loyalty programmes. To derive maximum benefits, one must broaden the thinking about CRM technology. The technology should be viewed as knowledge-based systems that seek to prioritize commitment of a firm's assets and resources to the more profitable customers while enhancing the relationships with ALL (right) customers. Pricing may not be the appropriate means to deal with the unprofitable customer. CRM systems give the firm access to information that may reveal other ways to manage the unprofitable customer. *Enabling* the managing of customer relationships is the goal of the CRM system.

Managing customer relationships is about selecting, acquiring, retaining, and enhancing relationships with customers by using an intimate knowledge of the customer's consumption domain to maximize the return on the firm's assets. With CRM systems, firms have the knowledge and the technological capability to identify, retain and enhance more desirable customer relationships.

This covers issues regarding which customers to acquire and retain for maximum sustainable profits. How do you assign value to customers and what is involved in that evaluation? Who is the right customer? Who should be in your customer portfolio? Firms look at customers as investments. How do you value these investments? Or, What is the equity of your customers?

THE RIGHT CUSTOMER

Markets consist of customers with diverse needs and differences in customer profitability. When serving multiple groups of customers, the goal must be to maximize the profitability of the combination of segments. When different segments are targeted, the firm essentially has a portfolio of customer segments — just as investors have an investment portfolio that maximizes returns at a certain level of risk, firms manage customer portfolios for maximum profits. As standard, hotels are fitted with an integral magnetic stripe reader for credit and debit cards and our EFT system is approved by all the major banks and clearing houses.

However, in an age when credit card fraud is on the increase, hotels can protect you and your customers from fraud with the addition of Chip and PIN devices. Chip and PIN solution can be integrated to operate with your Hotel billing system's configurable client credit limit.

In other words, this is the selection of customers at any point in time, compared to other segments, from which the firm can generate maximum profits. Firms must attract these most profitable customers and then must establish systems and procedures to retain them.

Most firms serve different segments depending on how the definition of segment is aggregated. The "segment-of-one" used in common business parlance to refer to customization at the individual level is really a maximum disaggregation of a market segment. As we saw, most firms facing fluctuating demand and experiencing peak, shoulder and low periods of demand find it inevitable that, at different times, different segments must be served.

Of course, each of these segments must have a compatible fit with the overall corporate image as well as a compatible fit with the products, services, employees and other customers. The question is, are they selected based on their combined long term value to the firm?

Most firms forecast the volume and revenues from various segments at certain price points at different times in the purchase cycle. In the typical firm, sales and communications efforts follow these underlying assumptions in pricing and in messages targeting sales prospects. The typical firm then attempts to formalize advertising campaigns to reach as many people as possible in the case of packaged goods. In the case of business-to-business services, the salespeople focus on making a sale.

Most firms then struggle to orchestrate all these messages for acquisition of the customer in a coordinated fashion. Much less attention is paid to the retention of the customer. Barring the well-run operation, common business practice for many firms is short-term oriented. On the contrary, a customer-focused firm begins by approaching sales as acquiring the right customer. The customer-focused firm also approaches the acquisition of customers as only the first step in managing the relationships with its customers. Once acquired, the customer must be retained. But, not always retained at any cost—a necessary condition is that the benefits outweigh the costs of acquiring *and* retaining that customer.

Customer-focused firms align all their activities and processes in acquiring and retaining the right customer. They see their value-creating assets as most profitably leveraged by focusing on the most valuable customers. In principle, the value of the customer is determined by contribution to the firm's objectives over the lifetime of the customer. The right customer is that customer whose inclusion in the firm's target market helps maximize returns on the firm's assets.

Customer Equity

Ask a manager, "What is your firm's most valuable asset?" Chances are that you will not get the response, "my customers." You will find even fewer firms that actually make an assessment of this value in any real sense. The valuation of the customer is implicit in sales figures — essentially, the revenues generated by customers of the firm. What is the flaw in using sales as a proxy in valuing a firm's customers? Consider this.

Two customers with the same cash value of purchases may not be of the same value to the firm. There may be differences in the cost of serving these two customers. The true value of the customer to the firm, or the equity that the firm has in that customer, must include all revenues and also all costs related to that customer, as in financial investments. A recent framework by a team of researchers defines customer equity as the "total of the discounted lifetime value of all the firm's customers."

They articulate three drivers of customer equity: value equity, brand equity, and relationship equity. They define value equity as the objective assessment of the utility of the brand, and value equity is driven by quality, price, and convenience; brand equity as the subjective assessment of the brand above and beyond the perceived value, and brand equity is driven by brand awareness, attitude toward the brand, and corporate ethics; and relationship equity as the tendency of the customer to stick with the brand, and relationship equity is driven by loyalty programmes, special recognition and treatment, affinity programmes, community-building programmes, and knowledge-building programmes.

Their framework — the customer equity diagnostic — is offered as a way to determine which customers to acquire as well as what will enable their retention. If we define customer value as the value of the firm or a product as perceived by the customer, customer equity is the converse and refers to the value of the customer to the firm. If the objective is to maximize the returns from your investments — the customer portfolio — you must maximize customer equity. This makes sense because you can maximize sustainable return on assets by maximizing customer equity for the long term.

To maximize customer equity, we must be able to measure it. If customer equity is simply the value of a customer to the firm, we can articulate that value in the same way that customers articulate our value to them. Just as customer value is the difference between benefits and costs to the customer, conversely, customer equity is the difference between the benefits and costs to the firm in serving a customer.

One team of researchers describes their method of measuring customer equity as follows: "[we] first measure each customer's expected contributions toward offsetting the company's fixed costs over the expected life of that customer. Then we discount the expected contributions to a net present value at the company's target rate of return for marketing investments. Finally, we add together the discounted expected contributions of all current customers." Evident in their metric are the following important points about customer equity:

- It is the sum of the equity of all of a firm's customers.
- It summates the gross contribution of each customer, taking into account benefits as well as costs of serving a customer.
- It includes consideration of future revenues and variable costs from each customer.
- It is a time-discounted present value of future benefits to the firm.

It becomes evident that the customer equity concept when disaggregated to the individual customer level allows us to look at the revenues and costs of serving an individual customer. It helps answer the question, which customer should we serve? The ideal customer to the firm is the one that gives it the maximum long-term customer equity, which theoretically you want as a brand-loyal customer. However, as we will see in the next section, the converse is not true, because not all brandloyal customers provide maximum customer equity.

WHO IS THE RIGHT CUSTOMER?

The concept of brand loyalty is a well-researched topic. The notion of a customer having a lifetime value and the prominence of database systems in managing customers have spawned a renewed interest in relationship marketing. Just making repurchases doesn't make a customer brand loyal. There must be some commitment by the customer to the firm for relational continuity reflecting a positive patronage bias. Not all brand-loyal customers have a positive disposition to the provider—some relationships may be forced because of a lack of choice.

Brand loyalty reflects a financial, social, or structural bonding with the customer. Relationship marketing seeks to enhance the mutual benefits from the relationship with the right customer—seeking brand loyalty from the right customer. To determine who the right customer is, we need to understand what benefits the firm expects from an ideal customer. Management system is designed with you in mind. Knowing the needs of modern hotels, simplifies the communication between the waiting and kitchen staff.

Customer's orders are communicated automatically to the kitchen for display on a kitchen management system or for printing on a kitchen printer. This minimises wait times and eradicates ordering mistakes that can irritate customers and cost your restaurant unearned revenue. When combined with handheld terminals, orders can be taken directly from tables, speeding up the ordering experience further.

At the outset, it is clear that the stream of purchases from the brandloyal customer is the primary benefit. In truth, the value of a brand-loyal customer to the firm must go beyond the purchases made over the lifetime of the relationship with that customer. As shown, revenues can be direct and indirect. Direct revenues are all of the customer's purchases, and indirect revenues are the cash value of all of the other benefits the customer accrues to the firm, in the form of direct revenues from referred customers. Costs to serve are drastically reduced, as brand-loyal customers are generally easier to serve. The argument is that as customers get familiar with the firm, its processes and products, customers will require less costly assistance from the firm in purchasing and using the product.

Loyal customers may even be able to open up ways in which to reduce the costs of serving the customer. Other benefits of brand loyalty would include product improvement contributions, new product opportunities, ideal sources for market information, and favourable word of mouth. At an overall level, all these benefits add up to a degree of stability and potential growth for the firms. Those brand-loyal customers who are committed in their patronage recognize their benefits from the firm and show commitment in the provider firm.

To select the right customer, the firm must be able to measure the value of the loyal customer from all these benefits as well as the costs of acquiring and retaining that customer. It is not easy to quantify all of these benefits. But it is possible to calculate the direct and indirect revenues from each customer. It is also possible to calculate the acquisition costs and relationship maintenance costs. Information technology has allowed us to obtain that data and made it easy to calculate the lifetime value of a customer.

The Calculus of Lifetime Value

A commonly used practice inherited from the direct marketing industry experience is called RFM (referring to "recency, frequency, and monetary data") — a method to determine whom to send promotions from among your customers. It takes into account the value of a customer's purchases, how recent they were, and how often they were purchased. As with the danger of any one approach, RFM has been used without much consideration for other important dimensions of the most valuable customer, such as how profitable the customer really is. RFM is therefore not a proxy measure for the lifetime value of a customer.

Indeed, if the costs of serving different customers show a variance, then the RFM method could attract the wrong customer. As, there are three main factors in calculating the lifetime value of a customer. It is not just a simple product of the value of each purchase and the number of times the customer will purchase the product over that customer's lifetime. The lifetime value of a customer should also include referral value. And, it should include the lifetime costs as well. CRM technology has made it possible to do this, but, how many firms actually use this understanding in how they design and implement their CRM solutions?

Lifetime Revenues

Lifetime revenues are the sum of all purchases that the customer will make. An assessment of the progression of purchases over the lifecycle of the customer is a key point to be made here. For instance, consider what a college student's financial lifecycle would mean to a bank. First, it is a savings or checking account with a debit card and perhaps even a credit card. The college student may also be a good candidate for an education loan. Once graduated, this ex-student is now in the market for a car loan, and quite soon a home mortgage.

Once other life events such as marriage and children occur, there are more car loans, mortgages, home equity loans, trusts, custody accounts, education loans for children, and associated financial products that a typical family would need. USAA, the life insurance and financial services company, follows marriages, births, and other life events so that it can advise customers on changing needs.

Every firm should attempt to develop such a long-term consumption profile for the typical customer in each segment, charting their potential purchases over a lifetime to take into account the life events of the customer—a customer lifecycle analysis. A similar case may be made for business customers based on an assessment of growth potential, so that any B2B firm has to project the growth of its customers and factor that into a lifetime value calculation.

Not all customers become brand loyal. We have seen that superior customer value is a prerequisite for brand loyalty. What percentages of new customers find the firm's customer value to be superior? The lifetime value of a customer must also factor the probability of the acquired customer becoming brand loyal. Thus, the lifetime value of a customer is the first purchase, plus the probability of repeat purchases for the duration of the relationship with the firm. The probability consideration takes into account that a customer may not be a good fit, that a competitor may have been able to provide a better fit, or that the need situation changed for the customer, such as in the case of relocation of the consuming unit.

Lifetime Costs

An activity-based costing approach allows a firm to account for direct costs that the firm incurs in the relationship with a specific customer. Once again, with information technology it is possible, where it makes sense, to attribute marketing and operating costs to individual customers. To obtain the full benefit from CRM investments, it is important to track costs at the individual customer level as well. Costs include acquiring and remarketing to the customer over the lifetime of the customer.

Costs also include value creation and delivery costs of serving the customer. Thus, the costs after acquisition include not only the costs of serving the customer, but also relationship maintenance and development costs such as the costs of cross-selling or upselling, called remarketing costs. Once again, the costs related to the first purchase are separated from that of lifetime purchases so that the probability of repeat purchase is taken into account when calculating lifetime costs.

Referral Value

The indirect revenue of referrals from a loyal customer is an often overlooked aspect of customer equity. The typical CRM solution and database marketing approaches ignore the referral power of a customer in calculating the lifetime value of a customer. Most firms do not even capture this data, partly because it is difficult to obtain information when knowledge systems in a firm are not configured to obtain it. The other reason may be that firms are not proactively looking for referrals from their loyal customers in any systematic way.

How is the value of referrals from a customer calculated? To answer this question requires looking at the process and mechanics of how referrals work. Who provides a referral? Customers who are satisfied and who have a certain degree of loyalty are the ones who are likely to convey favourable messages about a firm and its products and services. A key piece of information needed here is what level of satisfaction a brand-loyal customer needs to have before being likely to refer a customer. The next obvious question is, who receives the referral? Customers are likely to convey this information to family, friends, and colleagues.

However, not all who receive the referrals are appropriate customers for the firm. What proportion of the customers who received the referral are good candidates for the firm? Out of the ones that are appropriate for the firm, not all are likely to be suitably influenced by the referral to make the first purchase. Finally, if a purchase is actually made, what proportion of referred customers makes repeat purchases? Those that are moved to try the product may not all turn into loyal customers, but if they do, then their lifetime purchases add to the value of the customer making the initial referral.

MAXIMIZING LIFETIME VALUE

Maximizing the collective lifetime value of its customers should be the goal in managing customer relations. Head office Managers can analyse the performance of the business via the extensive reporting options on all aspects of your business and make product and price decisions for the whole organisation. Head office can either assume complete control of all matters (suitable for new sites or sites with inexperienced managers) or devolve varying amounts of control to site managers. At the furthest end of the spectrum you can reduce head office to a merely supervisory role where you only view global reports and produce financial audits. By increasing the number of customers *and* by increasing the lifetime value of each customer, firms can maximize their long-term profitability. To maximize the lifetime value of customers, you need to:

- Maximize lifetime revenues
- Maximize lifetime referrals
- Minimize lifetime costs

Maximizing Lifetime Revenues

Potential customers in a target market are by definition not yet customers. They may or may not be aware of or inclined to purchase from the firm. Firms have to nurture customers' dispositions to the firm through a number of stages before they can benefit from the loyalty of those customers. The acquisition process of customers could begin at any of the different stages on the road to loyalty. Customers may or not be aware or have an understanding of the features and benefits of the product. If the knowledge of the product is favourable, customers may have a liking or preference for the product.

Some customers may even have purchased the product but may not be repeat purchasers, may not be loyal to the product, or may have switched to a competitor. The right customers from any stage need to be moved toward a state of loyalty to the firm and its products. Can CRM technology be configured to capture information about the customer's stage in this process?

Once the appropriate customers are acquired, firms must look for increased usage and frequency, cross-selling opportunities, and trading up to premium versions. Retaining customers requires proactive customer management. If not, customers are likely to be indifferent to the firm, and a heavy user may become impatient and switch to the competition. Proactive customer management seeks to increase interest in the light user and treat the heavy user with appreciation and respect. Thus, the goal of maximizing lifetime revenues is achieved by maximizing the average purchase per year as well as the duration of the customer's relationship with the firm.

Maximizing Lifetime Referral Value

Referral customers have very low acquisition costs, since your brandloyal customers did the marketing for you. Following the mechanics of the process of word of mouth, maximizing lifetime referral value requires maximizing the number of referrals from each customer and maximizing the lifetime value of each referred customer. To maximize the number of referrals, one must maximize the drivers of referrals from a customer. In other words, firms must attempt to increase the probability of a customer making a referral of profitable customers. Firms must move loyal customers to become advocates of the firm.

When word of mouth is seen as a powerful customer acquisition tool, firms will find a way to stimulate referrals. Many firms are not proactive in stimulating referrals, assuming simply that "good word of mouth will happen if we do a good job." What must firms do to increase the value from referrals?

First, firms must identify the current and potential advocates of the firm. Firms must proactively look for individuals and organizations that are likely sources of referrals and stimulate and reward them for the right kind of referrals. Satisfied and loyal customers are not the only ones who can stimulate positive word of mouth. There are opinion leaders: such individuals and organizations as consumer reports, firms such as Gartner and Forrester, who might be considered experts in the field, and most important, the general and trade media.

Other ambassadors for the firm include players in often overlooked sources such as suppliers, resellers, employees, and other stakeholders such as investors. Community chat rooms on the Web are rich sources of information for the firm about what word of mouth is transpiring among its stakeholders.

Second, firms must attempt to stimulate referrals. Any new customer acquisition activity should record information on how the customer was moved to deal with the firm. When the firm obtains and records this information in the consumption profile of any customer, it is able to use it to calculate the average number of customer referrals made by each current customer of the firm. Customers and advocates with similar profiles need to be approached about playing a referral role. Since some customers may be amenable and others not, firms must explore what it would take to move suitable customers to play an advocate role.

Finally, from Equation C it is also clear that the better the fit between the referred customer and the firm's products, the greater the lifetime referral value. Research has also shown that the customers referred to by satisfied customers tend to be a particularly good fit with the firm and its products and services. The fit is better than with those customers attracted and acquired through marketing messages such as advertisements and coupon mailers. Brand-loyal advocates bring in like-minded customers. To get referrals to actually make a purchase or try the firm's offerings, it is necessary to get some incentives into the hands of the referred customers.

Minimizing Lifetime Costs

Reducing the costs of acquisition, costs of serving, and the cost of remarketing per customer can increase the lifetime value of a customer. Reducing costs to serve means that you need to be more efficient in the yield from your value-creating assets. We are drawn back to the yield management. Costs to serve per customer are reduced when you increase the cost efficiency or the capacity utilization efficiency. To reduce costs of acquisition and remarketing you need to be more effective in your marketing, both to new and to current customers. CRM technology can be used to continually learn from experience and utilize knowledge to reduce operating and marketing costs at the individual level.

REMARKETING TO THE RIGHT CUSTOMERS

To maximize the total lifetime value of their customers, firms must proactively manage customer relationships. As, proactive customer relationship management should seek customer appreciation from the heavy user and attract more interest from light users with better value bundles if they can contribute more in their lifetime value to the firm. This is in contrast to the firm that may have good customer service, for example, but in its reactive mode will leave the light user relatively indifferent to the firm and the heavy user susceptible to a competitor.

Most loyalty programmes involve some kind of reward for the continued patronage of their customers. Referred to as frequency marketing, continuity programmes, awards or points programmes, or simply loyalty a programme, the goal of these programmes is to market and manage relationships with customers. First instituted by American Airlines in the 1970s to get around government regulation that prevented price competition, now many firms' programmes attempt to solidify their relationships with its customers. The rewards or awards, as they are called, may come in the form of free products and services at the firm, or sometimes even free products from other firms.

The smart firms impose restrictions on the use of these rewards, so that they are used to increase demand when and where needed. For example, the hospitality industry blocks out certain dates so that the promotions are targeted at different segments to entice customer utilization at low demand periods. The problem is that many of these programmes are misguided in design and implementation. One obvious omission is evident when the lifetime calculus is compared to the criteria for the rewards. The rewards focus on revenues for the most part. That costs are taken into account is not clear at all.

A recent *Harvard Business Review* article offers data that would be a surprise for managers who have not considered costs of serving the customer. Brand-loyal customers are not necessarily profitable customers. In their knowledge of their own value and indispensability to the firm, the brand-loyal customer could make costly demands of the firm.

The firm, in doling out the goodies along with the product and other business favours, may actually be spending more than it recognizes as costs of serving that customer. The only criterion for these (brand) loyalty programmes should be that they should maximize the customer equity for each customer, based on the calculus of lifetime value that we have discussed. To be meaningful and powerful in managing relationships with the right customers, these customer acquisition and loyalty programmes must specify the following:

- Objectives of the programme
- Market and customer scope of the programme
- Value for customer scope of the programme
- Impact scope (costs, timing, impact on people, process, physical assets, other customers, publicity opportunities, etc.)

The objectives of the programme need to be stated in the context of customer relationship management. What is the expected value of the programme to the firm? What specifically must the programme accomplish? Perhaps it is to shift demand from peak to low periods, to attract new segments, to get light users to become heavy users, to promote a new product offering, or some other specific objective. Without a known objective, no programme can be assessed on merit.

The next specific decision for customer relationship programmes is to specify the customer that the programme is intended for. Often, frequency programmes are so loosely formulated that the rewards are earned by the wrong customer. Consider this example of the unfortunate consequence of a poorly guided programme. One Mr. Phillips logged 1.25 million frequent flier miles—about $25,000 worth of airline travel—with an investment of 50 hours of his time and $3,140. Mr. Phillips took advantage of an offer he found on the package of a Healthy Choice frozen entrée: 500 American Airlines miles for every 10 UPCs, with early birds receiving a double count of miles! A Mr. Fisher participated in the same programme and got 12,000 Northwest Airlines Miles.

Even if the airlines got an awfully good financial deal with Healthy Choice, one must ask who the loyalty programme attracts—whether the airlines are in the grocery business or in the airline business. How do their returns from this investment compare with the returns that they would have achieved with programmes that targeted the frequent flier instead of the frequent grocery shopper? The objectives of the programme should determine the appropriate segment or customer for the programme.

The problem with most programmes is that they don't take into account the profitability of the customer. Who should be the target of the programme? Not every customer is right for the firm, as we have seen, nor is every customer right for every programme. The best programme is one that tailors the award to the target customer. Thus, giant supermarket retailer UK-based Tesco mails 100,000 variations of promotions to its loyal customers.

Further, some customers may be more costly to retain, while others may be increasing their profitability to the firm. As, it is evident that firms need to look at the acquisition costs and retention costs of its customer base when determining the focus of their retention efforts. The calculus of the lifetime value of a customer is clear in its indication of what needs to be factored into deciding to which customer to target the promotion. Low acquisition costs and low retention costs are realized from your most profitable customers.

At the other end are your customers who were costly to acquire and cost you a lot to serve. Given the acquisition costs, the firm must keep a close eye on the customer's cost of retention. If the firm misjudged a customer's lifetime value or committed more resources to the acquisition of the customer than the revenue stream would justify, the firm has to either reduce retention costs or stop serving the customer if possible.

For the chosen customers, the firm needs to decide what they would do to motivate the customer to act. Customer knowledge should indicate what the customer is sensitive to, so that the value of the promotion can be appropriate for the specific customer or segment. The value of the promotion may be in the form of additional product, a complementary service, a straightforward price-break, or a discount for future purchases or a gift, among other things. Software firms are able to use lead customers in beta-tests and other benefits about new and innovative offerings before the rest of the market will find out, giving these B2B customers early mover advantage in their value chains in their industry.

Of course, a projected cost/benefit analysis is a must— what additional revenues have been realized and at what cost? For service firms, and where the programme is about service components, it is imperative that the promotion take into account what it would do to the demand patterns of customers. The last thing you want is for customers to strain your operation if your programme ends up drawing more customers during peak times when you are already at optimum levels of served customers. Any programme must consider the impact on other customers, on the staff, physical assets, and the delivery system or process.

For example, a shortage of airline seats infuriates frequent fliers who, as brand-loyal customers, having diligently set aside miles for use on family vacations, find they are not able to use the miles because use of awards is limited to a certain proportion of the volume on each flight and are restricted to certain times and dates. Estimates are that about 10 percent of all miles flown on a carrier can be from people cashing in on these rewards.

Does it make sense for frequent flier awards to range from personal digital assistants to designer watches? Are the costs and impact of these awards assessed? Another less obvious criterion is the potential for publicity in a promotion. A promotion that is newsworthy could attract the attention of a large number of potential customers and other stakeholders.

Firms must prioritize the management of customer relationships as a firm-wide imperative. Dell Computer, for example, has a "customer experience council" consisting of senior executives from each division or business line and major function that reports to a corporate vice chairman, no less. The council oversees measurement of several aspects of customer behaviour, including the effectiveness of its loyalty programmes. Dell even measures all the costs its customers incur in purchasing and using their products, including such things as shopping, ordering, installing, operating, servicing, and disposing of products.

The real value of tracking these revenues and costs over the lifetime value of the customer is that it allows firms to understand their brandloyal customers and anticipate their future needs. Such management practices will be able to deliver benefits to the heavy user to maximize their lifetime value. The lifetime value of the customer must be calculated for each customer — an assessment of revenues and costs over the lifetime of the customer's relationship with the firm. The investment in acquiring and retaining a customer must be made based on the customer's lifetime value to the firm.

A continuous monitoring of up-sell and cross-sell opportunities could increase the lifetime value of the customer. The potential customer equity from each segment or customer should determine the extent of value-creating adjustments to be made in terms of delivery process or product outcome customization in terms of value or other benefits to the customer or in terms of adjusting the price and other costs to the customer.

These changes in promotion offerings to acquire a new customer or remarket to a current or inactive customer will impact both costs and revenues and their effect on the lifetime value of the customer will determine subsequent investments in acquiring and retaining customers.

Guaranteeing Customer Value

Maximizing customer equity requires maximizing customer satisfaction by providing superior customer value. Ensuring product quality as the key driver of customer satisfaction requires checking the links between the firm's assessment of customer expectations and its ability to translate that assessment into product concept, operations design, and execution.

The doorman greeted the guest as the taxi pulled up to the Windsor Court Hotel, a 324-room hotel in downtown New Orleans, one of 120 independent luxury hotels of Preferred Hotels and Resorts Worldwide. Later, the waiter at the restaurant in the hotel accommodated off-the-menu orders — but the waiters did not make eye-contact with the diners. The guest room did not have the current edition of the Yellow Pages. This guest was one of Richey International's hotel spies, who had just conducted the Preferred test for the hotel. He has determined based on his experience as a customer that this hotel had met 88.5 percent of the Preferred standards, passing the 80 percent minimum. Firms such as Preferred want an unbiased assessment of the quality of their product and hire independent quality assessment firms such as Richey International to do the benchmarking of its hotels.

Firms also want to hear from you directly if you are a heavy user of their product. For instance, if you have over 1 million miles on your Sky Miles frequent flier account, or you fly 100,000 miles, or 100 trip segments, or 20 transoceanic segments in a year, you are a most-valued "platinum member" of Delta Air Lines and may be called to dinner in a private dining room in a luxurious setting. Delta wants to know what you think about the quality of their service.

Customers, especially the most profitable and valuable ones, are being called on for their perceptions of quality of the product received. Deliver a quality product and you are likely to have satisfied and loyal customers. The logic is pretty straightforward, and yet most firms struggle with at least some of the links in the chain: What is quality from the customer's perspective and how can the firm ensure the design and delivery of that quality?

Beginning in the late 1980s and continuing into the early 1990s, there was a groundswell of attention and interest in quality, primarily as a result of falling competitiveness of U.S. firms in the wake of Japan's advances. Quality gurus like

Deming, Juran, and Crosby preached the gospel of quality. The Total Quality Management (TQM) movement's cornerstone, the prestigious Baldrige award programme administered by the National Institute of Standards and Technology, was a coveted prize for a great deal of effort by companies large and small. *Business Week* ran a number of articles profiling such winners as L.L. Bean, FedEx, Xerox, Motorola, Disney, and others.

As established and emphasized throughout this book, an understanding of customer value is a fundamental element of customer focus and service orientation. This provides a model for delivering customer value by focusing on the quality of the product, as judged according to how it meets customer expectations. To maximize profitability, the firm must maximize the lifetime value of the customer.

A key driver of this goal is customer loyalty, which is dependent on customer satisfaction. To achieve customer loyalty, it is imperative that firms create and deliver a quality product—as defined by the customer. What does the customer evaluate in determining the quality of a product? And, what actions of the firm can be linked to that customer perception of quality?

Perhaps the most significant research on perceived quality in services was conducted by Parasuraman, Zeithaml, and Berry and sponsored by the Marketing Science Institute in Cambridge, Massachusetts. Their contribution was timely and thorough, and revealed a number of important facets of customer perceptions of service quality. These researchers defined service quality as the difference or gap between customer expectations and customer perceptions of service. Their systematic study of customer evaluations across a variety of services revealed that customers evaluate services along certain key dimensions grouped under the acronym RATER: the R eliability of the firm's products, the A ssurance customers feel that their needs and expectations will be met, the T angibles associated with the service, the E mpathy displayed by the firm, and the R esponsiveness of the firm to their specific and individual needs.

Further, they found that there were four principal factors that contributed to the difference between what customers expect and what the firm delivers:

- How accurate is the firm's understanding and interpretation of customer expectations?
- Is this understanding of customer expectations effectively translated into product design?
- What is the gap between what is envisioned and what is delivered?
- How good is the match between what is delivered and what is advertised?

Quality translates customer experience with the product to customer value. Quality of customer experience is framed with the costs of access to and use of the product. Customer value is what customers ultimately evaluate in making the decision on future patronage. Their loyalty to a solutions provider depends on their comparison of actual customer value with expected customer value. What determines how well a firm meets a customer's expected value?

THE MECHANICS AND MANAGEMENT

The model presented here is based on Parasuraman, Zeithaml, and Berry's work on service quality. It uses the same logic to incorporate more directly the notion of costs to the customer. *Quality* may not fully convey the point that it is customer value that drives the relationship with the customer. Analysis of quality without the associated costs will not completely capture the customer's frame of reference. The crux of the model is that there are an ideal and an actual for both the customer and the firm:

- The customer's ideal—expected customer value
- The customer's actual—experienced customer value

- The firm's ideal — product concept
- The firm's actual — operations design

The ideal and the actual of the customer's expectations and perceptions and the firm's product concept and operations design are placed in the context of the competition, the technology, the economy, and the legal and social environment within which they operate.

For example, competitive forces and the social environment can shape customer expectations and affect their comparative evaluation of customer value. The firm's ability to hire the appropriate personnel to deliver the product as conceived and designed can also be affected. Firms design a product concept based on their understanding of the customer's expectations, and then configure their value-creating assets accordingly in an operations design, which for the most part determines what they actually deliver as customer experience.

Thus, customer expectations, product concept, operations design, and customer experience are all linked. Customer value is dependent on this *expectations* link, the *design* link, and the *execution* link. The product concept needs to be closely aligned with customer expectations — the expectations link into the firm. The firm interprets and converts the customer's ideal — expected customer value — into what becomes the firm's ideal — product concept. The firm's actual — the operations design — is the firm's implementation of the firm's ideal — the product concept.

The design link is how well the firm has been able to implement the firm's concept of customer expectations with its productive assets. The design or configuration of a firm's assets executes the product concept, which is what the customer experiences. The execution link reflects how well the customer's actual experience is aligned to the intent of the operations design. Thus, firms must align the customer's ideal expectations to the product concept with a strong expectations link, which in turn is linked to the operations design by a strong design link, followed by an execution link, which determines the customer's experience.

Customer expectations management, studied with a sample of about 100 managers at the leading service firms in the U.K., was found to be composed of three key dimensions: keeping promises, marketing orientation, and employee skills. Another study in a small electric utility company in the United States found that frontline employees had a good understanding of customer expectations. If we assume that customer's expectations were perfectly understood, reflected in the product concept and translated into operations design and therefore customer experience, we should find that customer value is high.

However, in practice this is an invalid assumption. Customer expectations play a very critical role in shaping customers' perceptions of quality. When customer expectations are met, we find perceived quality is high, and when not met, we find perceived quality is low.

A closer look at the model will reveal that it is a good vehicle to frame all that can be done to improve customer value by increasing benefits and decreasing costs to the customer. It is a safe assumption to make that improving quality will improve customer value. Creators of the University of Michigan's American Customer Satisfaction Index have shown that customer expectations are linked to perceived value and customer satisfaction over time. Improving quality relative to costs to the customer requires strengthening all the links in the model. Tracking the logic from expectations to product concept, to operations design, to customer experience are a number of management actions that must be taken to ensure the desired customer value.

As, firms can make several check to ensure that the customer-focus and service orientation helps keep the value created and delivered closely aligned with customer expectations. Firms devote most of their attention to the value creation and delivery that affect the perceptions of the customer. However, if service quality is defined as the comparison between expectations and perceptions, shouldn't firms be focusing attention on shaping customer expectations as well?

Managing the Expectations Link

How closely the firm's conception of the product resembles the customer's expectations is a key conceptual link. This link is a function of how well the firm understands customer expectations and how well that understanding is translated into product concept and design. Evaluating the quality of products, especially that of services and service components of products, is difficult for the customer. To understand the implications of this difficulty, think of services as being dependent on search, experience, and credence dimensions. Some services are higher in search qualities that can be evaluated before the purchase. Experience qualities are those that can only be assessed during or after consumption, whereas credence qualities are hard to evaluate even after the service.

Similarly, customer expectations are just as complicated as customer evaluations. Research has shown that there is a zone of tolerance in customer expectations. The zone of tolerance is the range of expectations on any of the attributes that are important to the customer. The low end of the range is the minimum level of service that determines what is adequate. The high end of the range is the maximum level of service, referred to as what is desired. Customers will accept anything below the desired, as long it is above the adequate level.

To be able to manage expectations, firms need to understand how customer expectations are shaped. Firms need not only to understand customer expectations but to help shape them so that they are realistic. Customers form expectations from their own past experience with the firm's product as well as with similar products from competitors. In the absence of such experience, customers draw from their experience with the firm's other products or from similar products in general. The firm's own communications, both explicit and implicit, shape customer expectations.

Explicit communications come from all overt marketing messages. Implicitly, all visible actions and evidence of the firm—its prices, its location, its culture reflected in the way it conducts business—could communicate something about the firm. Customers are also influenced by others regarding consumption need situations and the solutions appropriate for those needs. This product-related information can come from messages they receive from the media, family, and friends. Firms are generally focused on their own overt communications and ignore the effect that other sources have on customers and their expectations.

It should become clear now that customer knowledge management is a critical starting point for any firm. The firm's effective use of CRM technology can enable the firm to have a more thorough understanding of customer expectations. Based on customer knowledge, we can see that a failure in this regard could be the result of ineffective use of available sources of information on the market segment and individual customer information. Are all the touch-points set up to capture and record customer information? Is there a process for sharing customer information across the firm?

For example, is there bottom-up communication from frontline employees to managers who participate in product design? Recognizing that customer expectations are dynamic, market research and customer research should be a continuous exercise. Instruments used to gather customer information should be evaluated for validity and reliability and should be easy to use. Inappropriate or inaccurate market research could result in management designing the product concept based on poor information from the marketplace about expected customer value.

An effective customer information system is critical in designing a product concept that would deliver the expected customer value. Management could perceive market needs and customer expectations inaccurately or make poor use of information in defining the service concept. How good is the information on the customer? How complete is it? How well has the information been analyzed, interpreted, and assessed by the product designers? Information regarding the market needs to be disseminated and shared among individuals who are responsible for the service design. Is the information current, and is there adequate indication of how these expectations are likely to change in the near and long-term future? Corresponding changes in product design should be prompt and responsive.

Linking the product concept to the needs/expectations of the customer is dependent on how well information has been obtained on customer expectations and how well they have been utilized in the product concept. The irony is that sometimes the information is about the wrong customer! The value and validity of the information for product design depends on whose expectations are obtained and used in developing the product concept. The heaviest users or, more precisely, the ones with the most lifetime value should be the priority. In other words, the weight given to the information should be prioritized by the lifetime value of the customer. There is also the possibility that the customer not being the right customer could simply be a result of poor segmentation methods or targeting the inappropriate segment(s) or customer(s).

As we have seen, it is important that the right customer be drawn to the firm. Marketing communications could be attracting customers who are not intended for the service. This might be manifested in poor media or message strategy, setting up inappropriate customer expectations for the product concept. The question here, of course, is whose expectations were obtained and who is actually using the product.

Thus, if the firm has not attracted the customer whose needs it can best satisfy at sustainable profits, then at the outset, the firm is at a disadvantage. Customer expectations are hard to measure. As we saw on customer knowledge management, firms may not have good information on the critical aspects of what customers expect. Research can be flawed and the data not valid, or the means of ascertaining the information is not reliable. Frequently the information is stale and customer expectations, being dynamic, have changed. If you have an inappropriate or inaccurate account of customer expectations, the product concept is not going to be aligned with what customers actually expect.

The Design Link

The link in the chain between what is conceived and what is designed into the firm and its value-creating processes depends on how effectively the firm's resources (people, raw materials, supplies, and systems) have been configured to create and deliver the product as per the product concept. The link could be broken with inaccurate configuration of a firm's resources, for instance, in poor selection of employees. In other words, the fit between job description (responsibility and tasks) and employee profile (skills, education, experience, and personality) represents a significant factor in how well the design is linked with the concept.

Poor empowerment of the frontline employee could fail the product concept in design. This could result from inadequate technical and customer service training, information systems, backroom efficiency, and other support structures. To be able to deliver the product as conceived or designed, internal marketing should be effective. For example, lack of clear communication of such things as the product concept, customer expectations, the marketing strategy, etc., to the frontline employee reflects poor internal marketing — and similarly in reverse, when the product concept decisions have not incorporated the frontline employee, the product is not conceived and operations not designed with that valuable input from the frontline.

In any firm, there is an inevitable disconnect between what the customers expect and what the frontline perceives that the customer wants or what the managers think that the customer wants. For the frontline personnel to be empowered, they require the knowledge and the tools as well as the authority and motivation to use them. If not, the firm will find a disconnect between product concept and execution.

The Execution Link

If the product concept is not supported by the appropriate corporate culture, frontline employees are at a disadvantage in executing the product concept. Backroom employees should see frontline employees as internal customers and the entire corporate culture focused on customer satisfaction. A number of things must happen for the firm to ensure that the product concept can be delivered. All productive factors must execute the design to deliver the product as conceived.

Employees need to be selected, hired, trained, and motivated to execute the design. They need to be empowered so that they have the ability and the inclination to provide the service as designed. Employees need to be allowed the discretion to accommodate and customize to each customer as far as possible. With the right kind of training, frontline employees in services other than just professional services can also make the right decisions.

What was actually delivered and the customer's perception of the product experience need to be continually monitored by the firm, so that the firm understands whether customer expectations were met. Perhaps, during customer interactions, the provider was ineffective in obtaining customer-specific information or in interpreting and using it in customizing the experience for each customer. In many services, customization is a key determinant in how well each customer is satisfied with the product experience.

For example, customers use tangibles (service provider's smile and other nonverbal cues) differently. Some customers need more assurance regarding the product than others. Some customers' information needs are different from others, and their evaluation of the service provider's responsiveness depends on how well the service provider is able to accommodate them. The service provider's ability to empathize with the customer reflects the ability of the service provider to understand the individual customer so that the actual delivery is closer to the individual customer.

Customer perceptions are just as critical as customer expectations. Customer experience may in reality be different from the intial concept. One way firms reduce the disparity between perceptions and reality is to encourage or require managers to work in the front line. By the same token, an argument can be made for allowing the frontline to work backstage, so that they are knowledgeable about how the backstage works to support the frontline.

Another way is to empower the frontline employees so that they have the flexibility and authority to accommodate specific customer preferences that standard operating procedures preclude. This is an effective customer relationship management practice as long as the front line is also appropriately trained to make informed judgments as to when and where such accommodations are reasonable and beneficial to the firm. Such a policy is especially useful to handle disgruntled customers following a product failure.

Promise only what you can Guarantee

Whatever resides in the minds of the managers as product concept is what is promised in advertising messages from the firm to the customer. And whatever resides in the minds of the frontline or sales people as product concept is promised during direct interactions with the customer. These promises are conceptual manifestations of the product. A useful diagnostic exercise is to check the customer experience as articulated in customer surveys and other methods of customer feedback against the product as imagined by managers and all customer contact personnel.

Similarly, the design of the operation in terms of the configuration of the productive factors can be checked agai it whatever resides in the minds of the customers as customer expectations. Is the firm's operation set up to deliver on those customer expectations? Customer expectations and the firm's promises must be delivered. Customer-focused firms should be able to guarantee what can be delivered. That which cannot be delivered must not be promised. Promise what can be delivered and deliver what is promised.

This means providing product performance such that customer perceptions are that expectations will be met. As we have also seen, every customer feels a certain amount of risk prior to purchase and during consumption. Even loyal customers know there is a risk of nonperformance and simply place confidence in the provider and are comfortable with taking that risk with that provider.

Helping with risk perception is a task of managing customers. Service guarantees are a powerful mechanism to manage customer's perceived risk in having their expectations met. A service guarantee is simply a promise to compensate the customer if service delivery fails established standards.

Expectations and Perceptions

For all the reasons inherent in the characteristics of services, intangibles in a product are bound to fail. There is no opportunity to "recall" a service, as is possible with the physical component of the value bundle. Therefore, guarantees are a meaningful and powerful technique to reduce the customer's perceived risk. The most important precondition is of course to ensure that the firm is able to deliver as promised, or in other words ensuring that you can deliver what you guarantee.

For example, Amtrak recently scrapped its service guarantee because it couldn't meet the goal of unconditional customer satisfaction or a complete refund of fare. They found the refunds were getting too expensive because of their frequency, about 4 per 1,000 passengers, which was four times what they had hoped for. In designing service guarantees, firms must consider the scope of the guarantee, the specific risks that the guarantees are aimed to alleviate, the ease of invoking a guarantee, etc. To make these decisions in a customer-focused manner requires an analysis of the customer's context.

What is it that needs to be guaranteed and from what? To get at this, one needs to understand the key customer concerns and what the perceived risk levers are. It is useful to examine this along the RATER dimensions since by definition these are the dimensions that customers use to evaluate quality. With this analysis, it is possible to determine what the scope of the guarantee should be. It is useful to think of the consumption activity cycle and the process blueprint to help with deciding the scope of what should be guaranteed.

Xerox and others in the copier business, for example, provide a guaranteed response time for service calls, or an "up-time" guarantee to relieve the risk of failure. Once the scope of the guarantee is decided, it now makes sense to think about under what conditions the guarantee is offered. This is a litmus test of sorts; if you cannot deliver something that you are guaranteeing, you tend to start putting conditions on the guarantee. The problem then is that the guarantee becomes meaningless. Conditional guarantees violate the spirit of a guarantee.

The fear of abuse of guarantees prompts firms to render guarantees worthless with numerous disclaimers and other conditions. The reality is that customer abuse of guarantees is minimal compared to the incremental benefits accrued by the firm from a guarantee. Customers who invoked guarantees cost Embassy Suites $3.94 million in 1996, while the hotel gained $23.14 million in incremental revenue from guests who said they stayed at the hotel only because of its guarantee.

At Hampton Inn it was $3.98 million in costs offset by $31.7 in incremental revenue. Following a significant loss of market share from poor quality, in the early 1980s Holiday Inn improved its standards and launched its successful satisfaction guaranteed programme: "Your room will be right when you check in and if not we will make it right or your stay is free." The conscientious firm will ensure that the quality is in place before promises are made in guarantees.

The benefits of service guarantees to the customer are obvious. The interesting thing about service guarantees is that there are benefits to employees as well. Team Xerox found that the employees knew what they were working toward and the service guarantees put the internal goals in perspective. The firm won its Malcolm Baldrige award because it was able to engage the whole firm in its quest for improving customer satisfaction by focusing on what customers were telling them about their firm and everything they did.

CUSTOMER SATISFACTION DATA

One of the most frustrating issues in practice is that most customer satisfaction measures and the data obtained are inept at guiding improvement efforts. A frequent reason for this is a basic flaw in the measures themselves. There is no way the firm can determine what a "courtesy rating" of 3 out of 5 means in terms of where and how improvement needs to be made. The way to get direction for product remedial and improvement actions from customer satisfaction data is to ask the question specifically so that you can trace customer perceptions to specific processes that pertain to customer experiences.

Take a look at surveys of customer satisfaction and ask yourself, does this information give me enough guidance on what I must do to improve customer satisfaction—or will I end up saying, "I know I have to improve frontline courtesy, but I don't know which touch-point the customer is referring to?" For actionable data from customers, it is imperative to align the customer feedback to specific actions or experiences that the customer has had with the firm.

A useful vehicle to sketch the customer experience and also a useful frame of reference for customer feedback on the quality of their experiences with the firm is a service blueprint. Ensuring that the customer survey questions are directed at specific points in the purchase and consumption experience requires aligning them to the blueprint, so that customers are relating their feedback to a specific point in any number of interactions they may have had with the firm, its products, its processes, or its people. The Xerox story about how it improved its processes to deliver quality is a case in point.

Xerox focused its whole organization around the goal of customer satisfaction and service quality.

It instituted the customer relations group at headquarters, regional, and district levels. It instituted the Customer Complaint Management System for improvements in technical services, information systems, and telephone system. Xerox developed tools to continuously measure, manage, and improve customer satisfaction. Two major sets of data were developed and utilized.

1. *External customer feedback data* included a series of customer satisfaction surveys as well as the Customer Complaint Management System. Four sets of surveys were used:

   • Periodic survey of a random sample of Xerox customers,

   • Post-installation survey of all Xerox Customers,

   • New product postinstallation survey of a random sample of customers,

2. Blind survey of Xerox and competitor's customers to establish benchmark levels.

3. *Internal quality and quantity measures of Xerox* included work processes and outputs that delivered products and services. Xerox processes that affected each area of customer interaction were identified and systems were put in place to measure and monitor these internal processes. The main objective was to provide leading indicators of Xerox performance and improvement opportunities.

Xerox mapped its customer feedback onto its (internal) value-creating and delivery process. This helped locate the necessary improvement effort on those processes that were responsible for customer comments.

The content in customer satisfaction surveys need to cover the RATER dimensions and to be mapped with the source attributable to each customer response on the survey. As standard, terminals are fitted with an integral magnetic stripe reader for credit and debit cards and system is approved by all the major banks and clearing houses. However, in an age when credit card fraud is on the increase, protect you and your customers from fraud with the addition of Chip and PIN devices.

Since the firm needs to prioritize allocation of its resources in any improvement to its value creation and delivery process, it would behoove the firm to understand what its most important customers say by cross-referencing the customer satisfaction data with the lifetime value of the customer. Ensuring that the value created is as close as possible to customer expectations is a daunting task — and perhaps the most important task for any firm.

Customer-focused firms will use feedback from customers and from the front line regarding customer experiences along the RATER dimensions. This customer information, when mapped onto the blueprint of the value-creating process, can identify the failpoints in the process. The processes at the failpoints need to be reviewed, redesigned, and monitored using customer and frontline feedback to ensure that customer value was improved.

The expectations link, the design link, and the execution link in the chain of intellectual and operational translations of customer expectations to customer experience will need to be closely examined to reveal where the links have been severed, causing customer value to be subpar. In an estate of hotels can have a varying level of control over operations. From full integration into your reservation system, through to company credit accounts, ledger accounts and room service, got it covered.

Chains of older hotels may offer a differing range of facilities, restaurant menus, price bands and wine lists, while modern chains are often custom built offering identical services at every outlet. Solution for hotels is flexible enough to meet the requirements of any estate of hotels. Working with Hotel industry can offer a complete end-to-end solution making complete hotel operation seamless.

In the hectic environment of a busy hotel, one of the main concerns for a hotel manager is that all the facilities, meals and drinks enjoyed by your customers are charged to the correct account and paid for before the customer leaves the hotel. Whereas previously customers might have checked out before the breakfast charge had made it onto the system at your front desk, that information is transferred in real-time. Whenever a customer charges a service to their room, system can interrogate the database to double-check the details of that occupant.

Chapter 4

Managing Customer Information

CUSTOMER INFORMATION

Firms cannot claim a customer focus or attempt to satisfy customers unless they have customer knowledge—that is, a deep understanding of the needs of the customer. With all the brouhaha regarding the growing importance of customer relationship management (CRM), too many hotel companies are still vexed when it comes down to delivering on a service promise. For hotel companies looking to differentiate themselves from the competition, exceeding guest expectations is paramount since it leads to building loyalty. In this complex era, it is crucial to truly understand and anticipate guest desires in order to fulfill that service promise.

But just how to do it effectively has been an elusive and vague notion not fully grasped by many hoteliers. Customer-focused firms really understand their customers by proactively acquiring information on the entire consumption process, especially the consumption activities of the customer. Such information must be processed and used to guide each and every interaction with that customer. This information can be used not only to make changes and improvements in the value-creation and delivery processes of a firm but also in managing relationships with the right customers.

Managerial decisions and actions cannot be customer oriented when customer data is lacking. Do we know enough about the customer's consumption activities to enable us to create and deliver customer-focused value and manage relationships with the customer? Hotel companies need to be in the customer intimacy business. They have to focus on delivering what the customer wants. You have to have some product leadership to deliver on customer intimacy but you have to focus on the relationship above all else. All goals should be focused on customers rather than the product or operations. And getting those loyal relationships are paramount to the property or brand's success. 10 percent of all hotel guests account for 44 percent of all hotel nights.

To get those loyal customers, it's crucial a hotel employ a culture of managing the "total customer experience" and also embrace a culture of managing "moments of truth" when that experience is on the line. Also, property leadership should demand a culture that puts the customer at the "centre of your world."

While a hotelier doesn't have direct control over what someone says after they walk out, it's very possible to control the guest experience in such a way that they leave with only positive thoughts. Customer intimacy is essential for success in today's world. "If people don't feel welcome and have an emotional connection, their loyalty will be in question."

It's also a matter of creating emotional loyalty rather than functional loyalty. Whereas functional loyalty comes into play if a hotel is conveniently located near an airport, for example, emotional loyalty transcends pure need and promotes frequent stays because the experience moves beyond the brand attributes to an attachment to the guest's psyche.

As Lew Platt, former CEO of Hewlett Packard, is widely quoted as saying: "If HP knew what HP knows, we would be three times as profitable." With the increasing ubiquity of information technology, there is an increasing focus on information—in a way, simply because it is there. Most managers understand the crucial role that good information plays in decision making. And here is a seemingly radical proposition: most data is underutilized in business. If you don't know what data is important, you may not recognize it when you see it. When there is an opportunity to obtain what may be valuable data, you will not access or gather, process, or utilize it.

Careful observation would reveal that as a norm, firms don't capitalize on opportunities to gather critical customer data. One reason may be that firms focus on the sort of data that is needed for a specific decision, such as in the development of a new product or an advertising campaign, especially in the B2C context, where there are no direct interactions with the customer. In the B2B context, where there is direct interaction with the customer, there is a greater amount of customer information being utilized for decision making. This is partly because formalized systems are in place to document customer information.

The customer being a business is very likely to have formalized purchasing procedures. Information-age technology has very naturally developed in the B2B case earlier and faster than in the B2C case. In the B2C business, where there is direct interaction with the customer, this technology is opening new thinking about customer data. A basic question is, what information is needed and for what purpose?

To manage the customer relationship, it's crucial to have deeper customer intelligence and then use that information to make each stay more memorable. For example, it is better to match up what your hotel does with each individual's lifestyle and give them something unique based on that lifestyle to reinforce the relationship. One suggestion made is to give, for example, a customer who stays 20 times a year and uses the gym regularly, a gym membership in their home town. That way, every time they go to the gym they will think of that hotel. It is a much better way to build loyalty then just giving them more points on their loyalty programme.

The Total Customer Experience. Customers want:

- A positive, enjoyable and unique experience from the beginning until the end of their experience

- They desire an emotional connection that, once made, often ensures long-term loyalty
- They must be valued for a lifelong contribution versus a one-time contribution to revenues
- Customers desire the perception that they are in control - ensure your customer-facing processes are engineered for how customers buy, when they buy and what they buy
- Customers want choice
- And they want this across all channels

The convergence of numerous information technologies in the gathering, storing, processing, disseminating, accessing, and using data has revolutionized the most generic of all business tasks — making informed business decisions. Communication, storage, and computational technologies have made this possible. Firms have found ways to deploy these technologies toward of increasing productivity. When the promise of technology is not delivered, as in many CRM and ERP projects, the focus has turned to how and why this data is useful for a business to support the investment in information technology.

One of the primary information systems of a firm is the customer information system. Customer-focused firms have used their information technology to develop smart customer information systems and broaden the scope of this business function. The role of the Chief Knowledge Officer, a title that was unheard of just a few years ago, is to manage the creation, discovery, and dissemination of knowledge in firms.

In the context of managing the knowledge resident with the firm about its business, customer information is one of the key pieces of the knowledge this person manages. The status given to this function is reflected in the seniority of the position and indicates the commitment of the firm to managing customer information that can be used to improve the value created and delivered by the firm.

The key issues discussed are: what kinds of customer information are needed to perform as a customer-focused firm, what information is required to usefully profile a customer, and how such a customer profile helps a firm understand the customer.

NEED OF CUSTOMER INFORMATION

The customer's desire to be known, confidence in service consistency, convenience, a feeling of welcoming and belonging and a comfort level where they know if there is a problem it will get handled right away. Of course, true loyalty goes beyond the physical hotel. These days loyalty is built with every single customer communication as well.

One way hotel companies communicate to customers is via emails. However, many companies are not communicating to their guests based on established preferences. For example, hotel brands will still offer ski vacations to die hard sun worshippers. At the hotel, many hotels are at the point where they are collecting information but are simply not tapping into that information to improve a guest stay. This can actually disenfranchise the guest since they have gone tot her trouble of providing their desires which are not being executed.

It's better to not gather the information, or say when it will actually be used. "It is more harmful to ask for that information and then not use it,In earlier, we saw that delivering superior customer value is a prerequisite for sustainable competitive advantage and that to be able to deliver superior customer value, a firm requires a good understanding of what customers value.

A good understanding of customer value presupposes good customer information. You cannot claim a customer focus, or seriously attempt to satisfy your customers, unless you understand their needs. An effective customer information system ensures that the appropriate knowledge of the customer is available for customer-focused decisions and actions. Without customer information, for example, it would be impossible to customize customer experiences, let alone manage customer relationships.

Knowledge of the customer is about being intimate with the customer. Customer intimacy is not just a cliché. Customer intimacy is a value discipline for firms that value and use an intimate knowledge of their customers to determine precisely which types of customers to serve and then align all their value-creation processes to create and deliver solutions to their specific type of customer. In a competitive landscape where competing products with little defensible product differentiation are becoming commodities, customization is one very powerful way to de-commoditize a product and build customer relationships.

Customization, as a term, is used broadly here to denote making changes in value creation and delivery such that the customer value is adjusted to each customer when and where appropriate. To know when and where customization is appropriate, you need to know your customer. Customer-focused firms pay attention to their customers. They listen to their customers. They really know their customers. The more knowledge the firm has about its customers, the more poised and able the firm is to deliver superior customer value.

Ritz-Carlton is known for its ability to personalize its service encounters with its frequent guests by maintaining a detailed preference profile at the individual customer level. With smart customer information systems, Ritz-Carlton's frontline employees are empowered to address guests by their first names and with knowledge of the guest's specific preferences. CLASS (Customer Loyalty Anticipation and Satisfaction) is an example of the information technology backbone to the Ritz-Carlton's guest relationships management programme.

Nadia Kyzer knows that having a database system is not enough, saying: "If our employees didn't put the information to real concrete use, the system would be worthless." Many firms use their knowledge of the customer to tailor their products and messages, or to promote new services to their customers.

For example, on the Internet, firms use knowledge gained from cookies to model patterns of customer behaviour on their Web sites so as to target promotions deemed appropriate for that customer. But good customer information management goes beyond targeting messages and promotions. QVC, the television-based retailer, uses a "daily activity report" to capture, monitor, and respond to customer perceptions of quality. Indeed, customer information should drive all business processes of the firm, including anticipating and meeting customer needs, creating and delivering services, and anticipating and recovering from service failure. Customer-focused firms must have good customer information to support the decisions and activities of the firm.

Customer-focused decisions affecting customer value require information on customers' needs, their product preference criteria leading to brand choice or loyalty and their perceptions and behaviour with regard to the product. This is true not only in the acquisition and use, but also in the disposition of products. Since service firms interact with the customer during consumption of the product there is ample opportunity to obtain and respond to customer information.

For example, New England's largest bank, FleetBoston, recently instituted the "customer experience crusade." The firm hired new staff to be stationed in the lobbies of its large-volume banks to serve as "meeter-greeters" assisting customers with their banking problems. Ostensibly, the attempt was to reverse the lousy service reputation FleetBoston had acquired due to its preoccupation with challenges following its merger with BankBoston.

With the knowledge gained from the activities of these customer service agents, the bank could be learning a lot about customer experiences and about how its customers actually use a number of the bank's services. This knowledge could then be used to drive all decisions about structuring and delivering the bank's services.

Good customer information is an asset of the firm. Logically, anything that contributes to sustainable competitive advantage is an asset to the firm. Since information on customers is useful in better matching a firm's products as solutions to customer needs, customer information is an asset that contributes to sustainable competitive advantage. For example, retailing has become so fiercely competitive that customer information assets are now seen as a competitive advantage.

Customer loyalty programmes in retailing are an attempt to link customer information to transaction data. Customer databases with purchase histories showing shopping preferences can be used in customercentric business decisions. Customer information is so valuable that firms who share customer information have to be careful about ownership of the database. At the end of a joint project, for example, Wells Fargo bank and Microsoft went to court to determine who owned a customer database that was created as part of that joint venture.

A good example of the power of customer information is on the Internet. Personalized Web pages use customer information "on-thefly" to customize the online customer experience. It is common knowledge now that firms have data that tells them who is visiting, how long they are staying on the site, and where on the site visitors are spending their time. Firms have knowledge of customer behaviour on their Web site with information on where visitors were entering the Web site (beyond the home page) and leaving the Web site, what keywords or links brought them to the site, and so on. All this is possible with data-warehousing and data-mining techniques.

In the Internet age, entire new services, such as online investment brokerage firms, have sprung up with value propositions enabled by technology-powered databases of customer information. In the brave new investment world, such new business models have forced traditional investment brokerage houses to incorporate these added value propositions to their existing businesses.

Other new financial service models have emerged. For example, a number of online firms offer instant mortgage or credit application and approval. Much of this is possible because rules-based algorithms created from studying databases allow firms first to offer conditional approval then then to be able to verify the credit risk because financial institutions have well-developed databases of an individual's financial history with them.

While the technology is available to do more with information, it is unfortunate that the kinds of information that most firms obtain on a regular basis are usually limited to transactional data—such as what was purchased, when, and for how much. With this information, firms are able to use recency, frequency, and monetary data and conduct what is called RFM analysis. Patterns of relationships between the variables are used to determine the best prospects among the customers in the database—those that might be high-propensity targets for a promotion. Transactional data is important, but not adequate to make customer-focused decisions.

In addition to being able to individualize interactions with the customer, with customer information that goes beyond transactional data a firm can derive knowledge to guide anything from cost-saving measures in operations to new product decisions. To really understand the customer, firms need information on a multitude of dimensions other than just transactions around their other actual consumption activities.

Kinds of Customer Information

Customer-focused decisions and actions require customer information at two levels. At the aggregate level, customer information is about the market—or more specifically, about the typical customer in the segment(s)—and at the individual level, it is about a specific customer. When the number of customers for a firm is large and there is no opportunity to directly interact with the customer, it does not make economic sense to get down to the level of individual customer-specific profiles.

For example, Procter & Gamble may not be able to interact with each and every Crest toothpaste user. In such a situation, the customer profile is aggregated and constructed at the market segment level. On the other hand, for firms that directly interact with the customer, it makes sense to profile specific individual customers. For example, General Electric interacts with individual customers of kitchen appliances who contract customer service agreements. GE can use their knowledge of each individual customer to target service contract renewal offers or offers for replacement products when the normal life of the appliance has passed.

In general, it should become apparent that the more the aggregation, the less specific and accurate the information on individual customers. The level of aggregation of the information will also determine whether that information is useful for individualized or segment-level decisions, processes, and activities. As the level of aggregation of customer information increases, the level of aggregation of the customer solution increases and the firm is working at the segment level. With lower levels of aggregation, the firm is working at the individual customer level.

Customer information, whether at the market segment level and/or at the individual level, provides a profile of the customer that guides a customer-focused firm's decisions and actions. A customer profile should be based on two broad categories of information: *Background profile* is general customer information not directly related to the product and is not product-specific; and *consumption profile* is directly related to the product and is product-specific, in that it is about behaviour related to the consumption of the specific product.

Defining Customer Identity

The background profile helps to identify the customer and to understand the customer's product-related behaviour. Firms have been diligent about gathering demographic and psychographic data to establish the background profiles of customers in the B2C market. Customer-focused firms use this background profile to understand product-related consumption behaviour. Such information is managed on an ongoing basis. In the B2C context, "who the customer is" can be profiled along the dimensions of demographic, psychographic, and consumption behaviour.

Of these three dimensions, the demographic and psychographic dimensions provide the background profile. Similarly, the background profile of the B2B customer needs to be in terms of their size, type of industry, their corporate culture, and other characteristics used to identify the firm.

In the B2C context, customers purchase and consume solutions for individual or household use. Consumers' consumption of products are determined to a great deal by their backgrounds. Their personality, family, economic, social, ethnic, and cultural profile influence the way they think, feel, and behave at home, with their friends and social circles and in society in general.

This background profile helps us understand their consumption profile and the customer value they seek in specific product solutions for their needs in the context of their activities. Similarly, in the B2B context, the background profile of customers helps us understand their consumption profile and the customer value they seek in product solutions in the context of their value chain activities.

Business customers purchase and consume solutions for their own business activities. The type of industry they operate in, what value they create, their size, the skills and knowledge sets of their employees, and the "personality" of their firm (their corporate culture) affect the way firms think, feel, and behave in the marketplace with all their stakeholders, especially with their suppliers and their customers.

The Consumption Profile

Whereas the background profile serves to identify the customer, to truly understand customer value one needs systematic information on the purchase and use of the product by the customer. This information is crucial to get a complete picture of benefits and costs to the customer. A consumption profile comprises information that provides a comprehensive view of the customer's knowledge, attitude, and behaviour with respect to a specific need and associated solutions. The consumption profile containing information on product-related behaviour helps profile a customer in terms of the how the product is acquired, consumed, and disposed. Such a framework allows the provider of the solution to get into the "customer activity cycle, " as Sandra Vandermerwe calls it.

Consumer behaviour models frequently divide a customer's consumption experience or customer activity cycle into the three stages of preconsumption, consumption, and postconsumption. In essence, when the customer recognizes a need to be filled, the customer

1. Processes the options and selects a solution to meet that need,

2. Engages in procuring and utilizing that solution, and

3. Evaluates the effectiveness of that solution.

The valence of the evaluation of the experience with the solution provider then determines the predisposition of the customer toward that solution provider for subsequent consumption cycles. A positive predisposition may materialize in the form of repurchase, brand loyalty, and favourable word-of-mouth behaviour. All of the customer activities in each of the three stages of consumption should be the focus of the firm's attention.

In some consumption contexts, these stages are not as distinct as in others. For example, in most B2B and high-involvement contexts, the three stages and their respective decisions and activities are quite pronounced, while in low-involvement consumption contexts these stages are not as pronounced and deliberate. Regardless of whether the customer is highly involved with the consumption context or not, the information available to the firm regarding the customer's behaviour (purchase, use, and disposition) should help align the firm's decisions and actions in all three stages of consumption.

It is important to note here the differences in the consumption of tangible-dominant products and intangible-dominant products vis-à-vis the firm's production activities. In the case of tangible-dominant products, the firm produces the product and then the customer consumes, or purchases and uses it. Whereas, with intangible-dominant products, the consumption process is temporally overlaid on the production process.

The Preconsumption Stage

Customer behaviour during this stage of the consumption cycle falls into three broad and sequential sets of activities: experiencing the need for a solution to a problem, identifying and analyzing the options to the problem, and deciding on a specific solution or way of meeting the need. For interactions with the customer during the preconsumption stage to be customer-focused, firms need to understand the perceptions, attitudes, and activities of customers as they go through their purchase decisions. Firms need to understand what customers expect from the solutions provider during this phase.

Since a product is a means to an end, to really know how the firm is doing with its products it is necessary to understand what the ends are. It becomes necessary to find out what role the firm's product plays in its use for the customer. It is also important to understand what those product roles might be in different usage situations. Often the customer's preferred solution can vary from one situation to another and yet the provider may not pay attention to that important change in the value that the customer is expecting from each customer experience.

The customer engages in accessing and assessing the relevant information on the ways in which a need can be met. Based on the information available to them, customers identify the various options and sort through information to determine the merits of each option. A number of complex issues about the customer experience in this stage are relevant for the firm. Where do customers get the information on the options — from what channels and in what form is the information sought?

What are the firms, brands, or solution providers that are evoked in the customer's mind? What are the criteria that the customer uses to evaluate the choice of providers of solutions? What are their expectations from the solution? How do customers benefit from the products and services?

When Nike introduced its slip-on sneaker Presto, teens could select designs for their shoes, attach matching music to them, and email them as music videos to friends. Nike was engaging itself in the preconsumption activities of the teen shopper. The firm acted on its understanding of how teens make and follow fads and fashion to introduce a new and innovative footwear design. If you can understand and reach into the customer's preconsumption activities, you can identify and attract the most appropriate customer for your firm's solution. Customer acquisition efforts can be more precise.

Customers make their choice after weighing all options against their criteria. Firms must learn how they make these vendor decisions. Who is involved in the purchase decision making? Who influences the customer's preferences and biases? Household (B2C) and organizational (B2B) consumption activities can involve several different individuals in a purchase decision.

Individuals play roles of users, buyers, influencers, decision makers, and so on, in what are called "buying centres." What are the various perspectives of each of these individuals and units engaged in the process? What are their respective roles in the decision? How is the purchase decision made, and what process do the customers follow? When and how often is the decision made? What are the preferences on the method of payment? How do these customers evaluate you and your products before purchase? Who is involved in that evaluation? This kind of detail in the purchasing activities of the customer, if gathered and utilized, can improve the effectiveness of the firm's customer acquisition efforts.

When the firm can interact directly with the customer, as in the case of services, there is the opportunity to obtain answers to these questions and respond immediately. Where there is an intermediary involved, the firm has to rely on the intermediary to represent the firm during this stage because the customer interactions are with the intermediary, not the firm. In the hand-held ("pocket PC" or "PDA") market, the competition is so fierce that companies are constantly monitoring the retailer to ensure that their products and promotions are well displayed and that the salespeople are well informed about them. The personal digital assistant maker Palm sends its agents as mystery shoppers to retailers carrying their products. Microsoft, for example, gave a free Casio EM500 Pocket PC device to salespeople who completed a Pocket PC training.

In first-time experiences with a service provider, customers may want to obtain information from the provider; thus that interaction is required prior to the consumption of the product. In situations where prices and terms of sale are negotiated, quite often the case in services, there is a significant amount of interaction. In most cases of intangible-dominant products, there is some customer involvement at least in the specification of the needs of the customer so that the specific solution can be configured for each customer.

Every firm should explore the opportunities it has for customer information to help in its acquisition efforts. The total solution includes the firm's contribution to the customer's preconsumption activities. Rich Hanks, Marriott's executive vice president of sales, recalls the early days of Marriott's Web presence. He says they did not even know that they had to put reservations on the first page. The way to design Web pages is to think like the customer and ask what the customer would like to do on the Web.

Whatever prepurchase decision-making process the customer will be using should be used in Web site design. For example, at what stage of the consumption activity cycle should a Web site charge a visitor for a service? Some visitors balk when they are asked to pay up front for a service they haven't used—they mightn't want to pay member fees when they only want one transaction, for example.

The Consumption Stage

Firms that are not customer focused are focused on product and sales. They typically pay a great deal of attention to the sales and purchase process — the preconsumption stage, but not the consumption process. During the consumption stage, it is important to understand how the customer uses the product or service and what kinds of customer activities are associated with the value that your firm's product provides. To truly understand customer value, the firm needs to dig deeper into the consumption activities of its customers. Who is involved in the use of the product? How do they use the product?

Does the customer have all of the information it needs on your firm's product to fully benefit from its use? Will the customer want to interact with the firm during the consumption stage? Will customers need customer service assistance, and if so, what kinds of usage assistance would they need? What kinds of issues dictate the effectiveness of the solution from the perspective of a consumption process?

Understanding when, where, and how a product is used is often a powerful piece of information if used effectively. Consider the following examples of firms using the context of the consumption activities as they relate to their products: Starbucks gives out free Frappuccino samples on a hot day at a busy square in Manhattan.

Unilever hands out free Lever 2000 handwipes at a food court in a shopping mall and body wash and deodorant to fitness class students at a Bally's Fitness Club. Schering-Plough provides blister treatment cushions to runners at a marathon. These firms are obtaining and using their knowledge of how their products are used at the point-of-use of the product. They are in a position to determine what customer education is necessary and what customer value improvement opportunities there may be.

Product-focused firms remain focused on the product and its sale and on not much after that. To these firms, most likely, a complaining customer is the cost of doing business and a necessary evil, and is frequently cast to the unwilling hands of a customer service person. Consider the example of the "easy to assemble yourself" products that come in a box with some assembly required by the consumer. The unsuspecting customer is frequently confronted by a confusing set of directions written by someone who doesn't need directions to assemble the product.

Procter & Gamble has been actively collecting first-hand information on how consumers use their packaged household products, just to understand how the customer uses the product—rather than depend on the pencil-and-paper survey method or focus group discussions. Those methods are fine for demographics and psychographics information, but are fraught with misinformation since they rely on the accuracy of memory and communication by the consumer as well as the interpretation and visualization on the part of the firm's researcher.

In fact, P&G goes as far as to send ethnographer filmmakers who spend full days—from the time a family wakes up until bedtime—for several days, just to understand how the firm can create value for the lives of families—a way to look for new product opportunities.

In the case of service-dominant products, since production and consumption are simultaneous, the consumption phase involves the firm's production activities as well. There could be a significant degree of customer interaction and participation in the production of the solution. The customer is part of the production process, and the information regarding each interaction needs to be carefully documented and utilized to deliver customer-focused experiences.

Customer roles need to be scripted and customer performances in these roles need to be monitored and managed. This information about customer interactions is critical in improving the processes involved in service delivery. Consider what package delivery staff have to do at some of their destinations. "Please place boxes in barrel. If they (it) don't fit, put on top of barrel—Dogs will rip up if left on ground, " is a set of instructions given to Land's End by a catalog customer for package delivery by a courier. Customer data must be collected, and the value creation and delivery system should be designed to obtain and use that information appropriately.

The Postconsumption Stage

Here, customer evaluations and their disposition to be loyal to your firm is the main feature of the customer's experience. When the product is predominantly a service product, the customer might be in a formal membership relationship with the service provider, such as in the case of banking or financial services, or enrolled in an educational programme, or a fitness club, or a contractual obligation, as in a B2B context.

At the end of each interaction in the case of service products, customers may assess their experience and determine their inclination toward the solution provider. In such cases, it would be important to collect customer reactions and impressions after each consumption experience so that the relationship is intact. With packaged goods, on the other hand, where there is no interaction during the use of the product, customer assessment of the provider is made during and after the use of the product.

What is the level of satisfaction with the features of your product and with the interactions they have had with your firm? They may be pointing out systematic failures in how you are creating and delivering customer solutions. Are there fail points in your value-creation processes that need more attention? It is possible to get some very critical process-improvement suggestions from the customer. How can the benefits to the customer be improved and costs to the customer reduced so that the net customer value is improved? Which aspects of your product are most and least useful for the customer?

This information would be useful in conducting a cost-benefit analysis of the service enhancements to your product. What is the value of each customer to your firm, by transaction and in the long-run? Can the value of the customer to the firm be increased? Are there opportunities to increase the role you are playing in meeting customer needs by providing a more complete solution either by adding products or adding features to the product? Are there ways to reduce the costs of serving the customer?

Although strictly speaking not part of the consumption profile, customers may have valuable information on how well your frontline personnel are providing value to the customer. Customers can provide you with an assessment of how well your intermediaries and resellers are representing you. For example, travel agents who represent hotels, airlines, and other travel-related services might need more information, attention or encouragement to represent you appropriately. All things being equal, retailers have no incentive to sell a particular brand in a nonexclusive retail store.

As long as the customer purchases the item from their store, the retailer does not care which particular brand is purchased. Customers are the best judge of your suppliers and vendors as well. If there are aspects of your value-creation processes that are subcontracted out, customer feedback is critical in assessing whether these contracts are working as desired and whether there need to be any changes in how these processes are structured.

How will the customer want to convey feedback? Have you provided the avenues that customers are most likely to use to tell you what they think about the solution and your performance? Are your touchpoints able to solicit, record, and report customer information? The next develops these issues around customer information management.

Understanding customer value is a prerequisite for being able to create and deliver superior customer solutions. Firms need to study customers' consumption profiles to understand their needs and what they seek in the best solution. The background profile of the customer is a useful context in which to analyze the information on the consumption activities of the customer. The consumption activity framework attempts to frame the information on the customer in the three stages of consumption activities. The types of information that would be required to help understand customer value.

Companies focus attention on attracting and acquiring customers so that data collection is generally geared toward the preconsumption phase. Existing customers' transaction data is used to guide attempts to generate more sales from them. When data on consumption and postconsumption activities are gathered, a great deal more can be done to inform the entire range of the firm's decisions and activities in creating and delivering the product.

CUSTOMER KNOWLEDGE MANAGEMENT

Among other reasons firms have a low return on their information technology investments is that they only poorly understand what the technology makes possible. The objectives and role of information in customer-focused management must be understood. Customer-intelligent firms use their customer knowledge to drive all of the firm's activities and decisions.

When firms have so much opportunity to interact with their customers, why is it that they treat their customers like strangers? The reason is not lack of information on their customers, but a lack in a firm's managing of that information. Firms may have the ability to capture that information, but may not process and utilize the knowledge to manage relationships with their customers. Information technology has only placed this realization in sharp focus.

On the other hand, the power of information technology has also raised the ugly specter of abuse of privacy when firms have access to a vast amount of customer information. Network Solutions, the first company that registered addresses on the .com, .org, and .net domains for the Internet, now a unit of VeriSign, raised eyebrows when it planned to sell names, street addresses, and other information gathered from businesses and individuals who signed up for an Internet address. There are several examples of Internet companies that have been reined in for their use of customer information. DoubleClick Inc., decided against combining Web-tracking data with offline databases. Toysmart was not allowed to sell its customer database when it shut down.

Hagel and Rayport suggest that customers will become savvy about their personal information and be reluctant to divulge it unless they receive some value for it. They predict that customers will win the battle over their information and firms they call "infomediaries" will "become custodians, agents and brokers of customer information, marketing it to businesses on consumers' behalf while protecting their privacy at the same time."

Nevertheless, customer information is crucial for a customer focus because of the need to make and execute customer-focused decisions in all aspects of the business. Firms will continue to require customer information to do business. Information technology will continue to challenge a firm's judgment in obtaining and using customer information.

Traditionally, research on the customer has been at the aggregate level, most utilized during the new product development process, to test new product ideas, in test marketing, and to test new campaign concepts in the realm of advertising. Marketing research has been standard practice preceding the launch of new products or advertising campaigns. Most firms conduct regular market research activities to direct segmentation decisions such as targeting and to determine how to reach segments.

Scholars have implored managers to "spend a day in the life of your customer" and to "get inside the lives of your customers, " so as to obtain a deeper understanding of customers. The quality of the knowledge on the customer is much greater because of direct observation. Obtaining that knowledge is only the first step. The basic skill of listening involves competence in sensing, evaluating *and* responding.

A study of 500 new car buyers found customer perceptions of these three dimensions of listening to be strongly linked to trust in the salesperson.

There are also benefits within the firm when the practice of listening to customers is emphasized by senior managers. Senior managers at First Chicago discovered that in emphasizing listening to the customer they found a cultural change within the organization. Customer knowledge gained from firsthand customer contact is very persuasive. Smart CEOs at firms like IBM, Cisco, and EMC intuitively realize this, and their senior managers contact their major customers on a regular basis. These firms also act on that information to improve relationships with those important customers.

The benefits of customer knowledge management have concentrated at the individual level. For example, Dreyfus, like other mutual fund firms, keeps track of client activity, claiming that they can predict when a client is going to shift money out of their mutual funds. Clients receive a call from a Dreyfus representative, who wants to know how the client feels about the investment and if his or her goals are being met. If Dreyfus is able to track clients at all stages of the consumption process, then the firm has a better chance of retaining the customer.

A number of CRM- and ERP-type software programmes allow companies to track processes in the life of a specific sales order from order entry all the way to delivery to customer. Of course, FedEx introducing package tracking is a well-known example from years ago. There are a number of benefits of information technology for firms of any size.

Siebel software allows Honeywell to spot problems and opportunities by tracking and analyzing all customer interactions; and at Marriott, Siebel empowers sales reps to respond more quickly to customer needs by integrating customer information from different departments. Lexmark International used Microstrategy software to help build what they call a data warehousing solution that accomplished a reduction in product delivery time by 70 percent, a threefold improvement in being able to deliver a customer's order on the desired date, and a 60 percent reduction in the information costs.

Package delivery firm UPS's selfservice tracking system, built by IBM, saved the company $450,000 a day in customer service expenses; and, the Clarify call centre software helped payroll-processing firm ADP improve customer retention rate by 5 percent and increase revenues by $100 million in 1999 simply by providing its customer service reps with customer information on its 8000 clients. There are numerous such examples of technology-based customer information management systems having a profound impact on a firm's productivity and in customer retention. Both outcomes are the basis for achieving and sustaining profits.

The marketing research practice has undergone a major transformation in recent years. A primary function of marketing research activities is to collect and analyze customer information. Technology has enabled all phases of the marketing research process ranging from sampling and data collection to analysis and reporting. With newer technologies and faster cycle times, firms conduct research at all stages of the product life cycle and not just at the product development stage or the test marketing stage.

This is at the aggregate level, whereas there is another change in research on the customer at the individual level, where technology is enabling customers to interact with the firm in a multitude of different ways not previously possible. These interactions can now occur during all stages of consumption. At every interaction with the customer, a firm has the opportunity to acquire valuable customer information, as well as to process and utilize the information for creating and delivering superior customer value.

Once the data is available, the challenge is to provide the user of the information and decision maker with targeted information. Enter rulesbased approaches, which are simply programmes that make sense of data using preconceived models. Sadly, the majority of CRM and ERP packages fail the firm not for lack of data, but simply because the firm does not have the appropriate understanding and appreciation of the fundamentals of customer-focused management. The rules for decision-making should be founded on customer-focused management principles.

The better the understanding of the customer and the utilization of that knowledge, the more customer focused the rules are. The best technology is only as good as what you tell it to do. CRM and ERP packages are really knowledge management software, tools to manage knowledge of the customer and enable the processes within the firm that are working to meet specific customer needs. To be more effective with these tools, firms need to gather the information, analyze and process it to obtain a good understanding of the customer, and utilize it at every opportunity in the value-creation and delivery process. Therefore, for a thorough understanding on how to manage customer information, it is important to understand the principles of knowledge management, in general.

Organizational Learning

What is knowledge management? To answer this question, we have first to deal with a more fundamental question: what is the difference between data, information, and knowledge? Are these terms just synonyms that are interchangeable? No, they are not! The simplest way to understand the differences among these terms is to picture a continuum. As we add value to data, it becomes information; and as we add value to information, information becomes knowledge. When we interpret data in a context, we convert that data into information that has some meaning. When we categorize and elaborate on that information with explanations, then we have added more value and the information can now be considered knowledge.

Thus, *knowledge management* is the management of data and information so that a firm can optimally capture, process, and utilize its knowledge for effective decision-making and operation. *Customer knowledge management* is the management of the acquisition, processing, and utilization of customer information for effective customer-focused decision making. Knowledge management practice has been studied in the context of how organizations learn. Before we discuss the methods of managing customer knowledge, it would be useful to understand knowledge management and, specifically, information processing from the perspective of organizational learning. Organizational learning can be broadly depicted as a three-stage process:

- Information acquisition or generation
- Information transmission or dissemination
- Information analysis or interpretation

Studies have demonstrated that a customer-focused firm has a formal and systematic method for gathering, interpreting and using customer information. To be customer-focused, therefore, firms need to be committed to the importance and value of customer information and to the use of customer information in guiding the creation and delivery of customer value. Thus, recognizing information's value to the firm and realizing when and where it is accessible is a prerequisite for knowledge management.

Richard Schulze, CEO of Bestbuy, the runaway success in electronics retailing, says: The best way to find out whether a store is running smoothly is by asking the cashier. "The cashier really understands where the problems lie. They're the last people to see the customer." To consider all of the sources of customer information we also need to first identify the natural state of the information. Customer information is available as tacit and as explicit knowledge.

The explicit knowledge is present in the databases of the firm or as documented information. Customer information files in databases are usually quantitative data. More and more firms are also using qualitative data when the technology is used to accommodate that type of data. Otherwise, most of this rich information remains as hidden knowledge. This information that is not recorded is the tacit knowledge that is usually resident with the employees and managers of the firm.

For example, records of customer service interactions capture the actual transaction, complaint, or feedback from the customer. At the same time, employees, based on their prior experience, will unconsciously record and bring to bear their judgments of the situation, which are not necessarily documented. The knowledge that employees have that is not recorded anywhere is deemed the tacit knowledge of the firm.

While at the individual level information is available from customer information files in the firm's databases, other types of information at the aggregate level may be gathered during the course of product development activities and other specific projects and reside in research reports. There is information on the customer at several locations within the firm, from sales to customer service, to name the most obvious areas.

Besides this explicit knowledge, there is a wealth of information at both the aggregate and the individual levels. Wherever the firm has interfaced with the customer, the "touch-points, " there is information on the customer. It immediately becomes apparent that several product and service delivery contact points within the firm have access to customer information, but not all are part of the customer information management system.

CRM software packages enable the firm to capture all of the information gathered at each touch-point into a central database and make it available at any touch point. To determine all of the opportunities for gathering information on the customer, firms can develop a blueprint of the order delivery or customer service processes depicting all of the touch-points from the consumption cycle when the customer interacts with the firm.

Knowledge gained on the customer from each and every customer interaction when recorded and made available can be valuable in improving the customer focus of any of the firm's decisions or actions. The major sources for customer knowledge are discussed below.

Customer Information Files are typically transaction-related and purchase-related individual-level customer data found in sales records, customer interaction records, and customer service incident records. What is perhaps not in these files, which needs to be considered as a necessary component of individual customer profiles where available, is information related to the activities of preconsumption, consumption, and postconsumption.

Research Reports have segment-level aggregate data that are found in product development reports, advertising research reports, sales and marketing reports, and customer service reports. What needs to be seen is whether research on market segments is an ongoing pursuit of the firm — to indicate that the firm is proactively assessing customer value being delivered by the firm.

Knowledge held by employees and managers — Tacit knowledge of the customers — is by definition not completely identified and its utility not realized. If there is no formal process to regularly capture and utilize this information, a wealth of knowledge resident in the firm is being ignored.

Upstream and downstream sources: When suppliers and intermediaries are regularly included in discussions of value-creation and delivery decisions, there is an opportunity to discover information on market trends, technology, ideas for product improvement, and so on that are useful at the aggregate level. At the individual level, where there are (downstream) intermediaries, if it is possible, it would be highly valuable to obtain individual customer level information.

Now let us look at opportunities that may be missed by firms that are not proactive in looking for customer information. For example, how many firms capture and utilize data on referrals? If it is appropriately captured and analyzed, firms can use this information to proactively stimulate referrals. Firms can identify customers who could be bringing in new customers. Firms can determine what incentives would be effective in stimulating referrals. From the referring and referred customers, firms can learn what messages are being conveyed about the firm and which messages are most effective in stimulating product trial from new customers.

Firms can learn what types of customers are most liable to switch from their current patronage based on word of mouth. Firms can track and follow the patronage behaviour of these new customers and nurture their loyalty and referral behaviour. Firms can also find out from these new customers valuable information on the competition as well. Similarly, what about prospective customer queries? How well is this information captured and utilized? What would such information contain?

They are a good source of what customers look for in a product, a reflection of their choice criteria. How many firms systematically record information on complaints and praises? This type of information would be very helpful in understanding how the firm is faring on creating and delivering customer value. It could highlight failpoints in the value-creating processes and provide input into frontline evaluation and motivation activities.

Many firms have loyalty programmes to reward the frequent user. What sort of frequent-user data do these firms maintain? How many firms really use the frequent-purchase data? Some firms who know their customers well recognize the importance of sorting out the most profitable customers from their customer base and treating them differently. Some customers are indeed more equal than others! Most firms use frequent-purchaser data for targeting specials and promotions. However, this data can also provide invaluable information on customer value and customer profile information to align the firm's value-creation and delivery processes.

Traditionally, firms have gathered routine customer information only at purchase and when a customer returns a product for a refund or invokes a guarantee. Firms have also attempted to obtain customer reactions through customer feedback mechanisms. Feedback forms and customer surveys are administered on a regular basis in some firms. However, most often the information gathered from these efforts is worthless because of faulty designs and worse, sometimes even result in misdirected decisions because of flawed survey methodology. More often than not, the surveys are neither reliable nor valid.

While reliability and validity as statistical measures of the psychometric stability of a scale are important for a scientific, unbiased study, management can improve the utility of customer surveys with a rather simple analysis. Examine the design, administration, and use of any survey along the dimensions of structure, content, and process. Structural features such as length and the time it would take the respondent to fill out the survey, the layout of the various questions in the survey, and the response format for the questions affect the effectiveness of the survey.

Content decisions, such as the categories of information sought from the respondent and information identifying the type of respondent filling out the survey, affect the utility of the survey, which will only be as good as the effectiveness of its administration and its usability. Thus, the process of survey delivery, customer response delivery mechanisms, and the customization of the reports to the decision maker and user of the information in their value-creating and delivery roles is critical to the success of the customer feedback exercise.

There are other sources of customer information that the firm has access to and may not fully utilize. The systematic way of determining what these sources might be is to list all the possible touch-points where customers interact with the firm and placing them in sequence in the three phases of consumption. In the preconsumption phase, customers may interact with salespeople or customer service or other frontline personnel.

Recognizing what these touch-points are and what kinds of information is shared between the firm and the customer is an important step. Firms must decide what data needs to be captured at the touch-points and how it is to be used in guiding decisions and activities.

How many firms systematically utilize frontline operations personnel for gathering customer information? Frontline operations personnel are in an ideal position to gather customer information. They can be trained on when, what, and how to observe or elicit information from customers. They need to be motivated and recognized for their ability to obtain, process, and disseminate such information.

Similarly, there are also the intermediaries or resellers of the firm's products and services that represent the firm in different ways. These intermediaries might perform functions of delivery and distribution, promotion, or customer service on behalf of.the firm. Walmart recognized and wielded the power it held with such customer information in negotiating terms with powerful packaged goods manufacturers, including Procter & Gamble.

In the consumption phase, customers may call for assistance in using the product or on some aspect of how the solution works. This information needs to be captured and incorporated into customer education activities such as product directions-for-use instructions, or used in product development and improvement efforts. There is also a wealth of information within the firm available from customer complaints and suggestions.

Service failures and recovery situations provide valuable information on customers and their experiences with the firm and its solutions. If these are recorded and maintained on a systematic basis, they can be "mined" for patterns that yield insights not otherwise available. For example, recurrent complaints about customer service response time might suggest that the customer service process needs redesigning.

Often when the data is not available on which to base a decision, the firm has to collect that information. For example, say the firm has received several complaints about customer service access and wants to determine what changes need to be made to the customer service operation. In this case, the firm may want to conduct some research in order to determine the customer needs and preferences with regard to customer service.

For each decision that directly relates to a customer, the firm has first to determine what kind of information is needed to make the most effective customer-focused decision. Some of this information is available in existing databases, while some may have to be gathered. To gather information, the firm needs to set up a method of determining what information needs to be gathered, how it can be obtained, and in what form it will be available.

The Research Process

When the information is not available from the day-to-day activities of the firm, then research needs to be conducted on the customer on a project-by-project basis. There are a number of choices and decisions to be made regarding the methods available in researching the customer. For example, to determine what changes are needed to improve customer satisfaction, there are formal and informal customer satisfaction measures. There are qualitative and quantitative methods to obtain the information.

For unique project-based decisions, there are several datacollection methods. Among the quantitative methods typically used are paper and pencil surveys administered over the telephone or by mail. On the other hand, qualitative methods typically utilize structured and unstructured interviews, focus groups, or observation methods. A number of innovative qualitative approaches are being introduced.

For example, Kimberly Clark for its diaper market, Intuit with its personal finance software, and Patagonia for its outdoor gear have used the method of "story-telling" by getting customers to tell real-life stories expressing how they use the product and feel about it. Customer case research is another method that conducts chronological case studies of actual purchases to discover previously unknown purchase drivers by systematically capturing all events leading up to the purchase decision. These approaches can uncover customer experiences and thought processes that may not be accessible through a pencil-and-paper survey approach.

Any research project should take a systematic approach for maximum benefit and to reduce errors (such as measurement and sampling errors) and biases. As depicted, in general, the steps are:

- *Framing the research* — Management needs for the decision or objective.
- *Examining existing and available information* — for hypotheses, specific research questions and to guide research design.
- Specifying the research questions.
- *Determining the research design* — Questionnaire construction, sampling frame, data collection method, and plan of analysis.
- Administering the research and collecting and analyzing the data.
- Reporting the results and findings.

First, the research objective frames the management decision to be made. Information already available is used to guide the research design. The research is then designed by making decisions on what information is needed and how and where to collect the information. The research design includes determining the questions to solicit the information, the method for administering these questions, the sample of subjects, and a plan of analysis. Qualitative research designs are appropriate when there is a need to generate ideas on what to focus on, or to get in-depth information on specific issues. Qualitative research, such as focus-group methods, can complement quantitative research, such as surveys.

Quantitative research designs are useful when a broader cross-section of the customers needs to be polled about a broad range of issues. Sampling techniques are used when it is not possible or necessary to contact all the sources of information. Statistical techniques allow for generalizing inferences and conclusions on the entire population of interest based on data from a representative sample of that population. The research instrument or questions and response format are constructed and pretested on a small subset of the population for validity and reliability before obtaining responses from the final sample.

The responses are tabulated and statistical procedures applied to the data for analysis. The findings generated are then made available to guide the decision maker. The appropriate format of tables and the report are predetermined in the plan of analysis, so that the data is collected and analyzed according to the needs of the user of that information.

Interpretation, and Utilization

Especially in service firms and in the service component of manufacturing firms, where there is a significant opportunity to interact with the customer and to customize value, firms have used customer information in a number of ways. Observe any service encounter and you will see tacit or explicit customer knowledge being used. When the data from these interactions are systematically documented and analyzed, patterns in customer data can detect the dynamics in consumption behaviour. When aggregated, customer data provides an opportunity to analyze customer segments for anything from price sensitivity to demand patterns to why, when, and how customers behave in the various activities within each of the consumption stages.

Firms can divide their customer base into segments based on profitability. When customer purchase information is mapped onto capacity utilization data, for example, a deeper analysis of the firm's value-creation and delivery processes can help determine which of the firm's marketing and operations practices contribute to its profitability. Which frequent customer promotion programmes are most/least effective? Where are the low and peak demand periods? A yield analysis of the utilization levels at the various price levels would help determine the most effective pricing decisions.

Each customer has a value to the firm, and that information is also valuable to the firm to determine the right customer. The lifetime value of the customer to the firm is a function of what they bring to the firm as revenues over the lifetime of the customer. By the same token, it is also important to know what it costs the firm to serve that customer. Loyal customers with a favourable disposition toward the service provider are also likely to engage in favourable word-of-mouth behaviour. They are valuable in trying new products being offered by the firm.

They are also less costly to serve because they are familiar with the firm and are socialized into the procedures of the firm that they encounter in their consumption. As customers' loyalty with the firm grows over time, the firm also has a better knowledge of the customer and is more efficient and effective in serving that customer.

Firms that have the data on their customers are able to engage their best resources for their best customers. For example, firms have used information on a customer's profitability to the firm to determine the level of service appropriate for each customer. A *Business Week* (October 23, 2000) special report titled "Why Service Stinks" cites several examples of firms pampering their premium customers. Charles Schwab's premium customer waits no longer than 15 seconds for a phone call to be answered, while others could wait as long as 10 minutes.

Centura Bank's most profitable customers — rated 5 on a scale of 1 to 5 — get an annual call from the CEO; customer retention in this group improved by 50 percent in five years. Even utility companies, believed to be less sophisticated in their marketing than the consumer packaged goods firms, have a higher proportion of service representatives per customer for their higher-revenue customer groups. In each example, unless the firm knew what each customer was worth to the firm, it would not be able to proportionally allocate its resources to the "right" customers.

A grid guided by the three consumption stages can be a useful framework for determining what decisions need to be based on what customer information. All the functions and processes that affect the creation and delivery of customer value in each of the stages need to be aligned with customer information.

TECHNOLOGY-ENABLED CUSTOMER INFORMATION MANAGEMENT

Technology has revolutionized customer information management. Information technologies allow firms to gather, store, and interpret demographic and behavioural data so that each subsequent interaction with the customer is customized, based on historical data gathered from each customer interaction. Digitized purchase transactions in commerce have enabled numerous opportunities to track customer interactions. Chip technology has progressed so rapidly that soon there could be a computer chip in just about any product that would enable mindboggling links to improve customer value and the customer-firm relationship.

For example, take the Internet and e-commerce. Most Web sites will ask simple unobtrusive questions of customers when they browse the site. This information identifies the customer and tracks the customer's activities or "clickstreams" on the Web site. Subsequently, software can use this information to personalize the Web site and the experience for the customer.

A number of firms have emerged that provide customer information management software. Such software can customize Web pages for each customer depending on the customer's profile, which is created based on the customer's previous interactions with the firm and the Web site. Vignette's software allows Web sites to dynamically generate pages that are custom-made for each customer. Silknet's virtual sales assistant guides customers through a series of questions about their need situations and then makes product recommendations.

Marriott has an enterprise-wide software that integrates information on customers from different departments so that its sales reps can anticipate and respond more quickly to customer needs. Similarly, H&R Block uses software from Clarify to combine and coordinate customer records from its tax offices, discount brokerages, its customer service centre, and its Web site so that customer interaction personnel in each department have access to all of the information on the customer.

There are a variety of methods for analyzing data. Data mining techniques attempt to discern patterns in the data where one set of characteristics affects another set in a certain way. SAS Enterprise Dataminer software helps detect patterns in databases and construct models to predict customer response to the firm's actions. Software from iMarket looks at customers' most recent or most frequent purchase decisions and the monetary value of these purchases for the firm and helps decision makers select a subset of customers for targeting specific promotions or new product introductions. Epiphany and Digimine are examples of firms that use a set of their own proprietary algorithms to analyze customer data for their clients.

Early clients of Epiphany included Charles Schwab and Hewlett Packard, and for Digimine include Nordstrom. Edify has software that will link customer feedback such as survey responses to individual customer profiles for anlaysis. There is also software such as Nudist available for the analysis of qualitative data. When firms are able to sort through their customer databases and group customers by the relevant characteristics with such datamining software, the power of segmentation and targeting is fully realized with such datamining software.

In general, data is useless for decision making unless it is organized for interpretation. When organized appropriately, data can be contextualized so as to be transformed into information. Displaying and disseminating information is critical to the firm's success. Without proper reporting procedures, decision makers do not benefit from the information that the firm has.

For example, ESRI has a programme called Arcview that allows the user to spatially display numerical data. The user can view the data in his or her preferred format to match individual cognitive styles. As Peter Drucker notes, no two executives organize the same information in the same way. Therefore, software that allows the users to display information in their own preferred formats is valuable in customer information management.

The task of the customer information management system is to generate intelligence on the customer so that firm can be more customer-focused in its serving the customer. Intelligence generation capability has been shown to be positively correlated to superior customer value. In a study that involved hour-long in-depth interviews and surveys of sixty-six of the largest one thousand electronic firms by scholars interested in the characteristics of the market-oriented firm, firms are urged to treat learning about the customer as an investment.

Ultimately, customer information is useless unless it is available to the user. Scholars researching the value of information have called this the "value in use" metric. The more the information is used, the more value it has! Customer information management, therefore, involves determining how the information needs to be made available, to whom, when, and in what form. The focus here is on who needs the information.

In the spirit of customer focus, the information system serves an internal customer—the decision maker. What will the user need the information for? What specific information is needed for the firm's decisions and activities? Is the user of the information able to enhance customer value with the information?

At what level of aggregation is the information used: policy level and customer level, segment level, and individual level? Such decisions need to be made so as to prioritize the customer information according to its impact on customer value. In making judgments about committing resources to managing and leveraging customer information, the firm should consider the worth of the information in terms of how it would help the firm provide superior customer value at a reasonable profit.

Customer information must drive any business process and activity. Customer-focused management requires a customer-focused analysis of the information necessary for designing processes and making decisions regarding value creation in the firm. These run the gamut of decisions regarding creating and delivering customer solutions, product improvements, attracting and retaining customers, customer service, recovering from product failures, hiring and motivating employees, and ensuring a customer-oriented climate and culture.

Creating and Delivering Superior Customer Value

Firms create and deliver customer value as a bundle of benefits contained in a core and supplementary product. Since the core of any product is a commodity, the competitive advantage comes from the supplementary product. Designing and delivering superior customer value is all about adding value to the customer's value chain and doing it better than the competition. Paying careful attention to the details in each and every encounter with the customer ensures a customer-focused delivery of the value created by the firm.

Since the customer-focused firm competes on service, it also faces the classic challenge of any service firm—matching demand and supply. In order to maximize the profitability from its value-creating assets, firms have to determine the most profitable mix of segments for its customer portfolio.

Designing and Developing Superior Customer Value

How do you design customer-focused solutions that compete on service? Let customer information drive the decision on how best to leverage the productive assets of the firm. Market-based as well as asset-based assessments are necessary to configure value-creating assets into core and supplementary product benefits that provide a sustainable competitive advantage.

How does a small startup bank beat the big boys of the banking industry in their own backyard? Commerce Bancorp in New Jersey is not your standard bank. You will find their branches open at 8:00 AM and at 8:00 PM, and for most of the day on weekends. Vernon Hill, founding CEO of the bank, who also owns Burger King franchises, has embraced retailing practices in successful firms such as Walmart, Home Depot, and Starbuck's. The core product at Commerce Bancorp is the business of banking, of course. But the total solution includes access and convenience benefits for the customer that most definitely distinguish it from other banks.

There are even bathrooms for patrons in each branch at Commerce Bancorp. The total solution for any product includes a core product with supplementary features that in combination present a resultant value to the customer. Product differentiation and superior customer value is not in its core product, but in that total solution. The total solution for Commerce Bancorp is not just banking delivered by human or automated tellers in physical branches, it is banking at the customer's convenience when, where, and how the customer wants it.

Indeed, the difference in a customer-oriented firm is very evident at any Commerce Bancorp branch. Commerce bank has found powerful ways of differentiating itself by being customer focused and then it has been able to execute its customer-focused value creation and delivery.

Firms create and deliver value from their assets. Firms must determine how they might be able to create customer value that is superior to the alternatives available to the customer. With a customer focus, the firm seeks to design a superior solution that will provide the firm with sustainable profits. This covers issues and decisions involved in conceiving, designing, and delivering the customer-focused product offering. What must happen to ensure that the productive assets of the firm, facilities, processes, and people, are set up to create and deliver customer-focused value?

TRANSLATING CUSTOMER NEEDS TO PRODUCT CONCEPT

The assessment must be done both at the aggregate, segment level and at the individual, customer level, where appropriate. The aggregate, segment-level assessment orients the firm's thinking about overall product strategy. When a firm directly interacts with its customer, the individual, customer-level assessment helps the firm configure the product within that strategy to the individual customer.

In keeping with the view that a product is really a bundle of benefits, the core value in any product for any firm is the primary benefit that the firm provides. The supplementary product includes as parts of the solution additional benefits that augment the core product.

Typically, a large part of the supplementary product is comprised of services. For the most part, the core product is a commodity, as in the example of Commerce Bancorp. All banks offer deposits, withdrawals, and investments and banking products such as checking and savings accounts, and fixed and variable income securities, and today most banks offer mutual funds. These products comprise the core product for banks.

How these products are different from bank to bank is represented in the supplementary components of the total product. Commerce Bancorp offers free checking and money orders and does not compete on its lower interest on its savings accounts and CDs. Thus, Commerce Bancorp differentiates itself on a number of different supplementary benefits to the customer, ranging from the quality of its customer interactions to a number of facilitating services such as bathrooms in its branches. Similarly, most if not all banks have an online presence, but customers at banks like Allfirst Financial can talk with bank personnel via their home personal computer.

The hotel industry is one that grew out of its room-and-board days to a total product offering that today includes a multitude of facilitating services. "Room and board" was the commodity. As all hotels in a certain level, competitors in the same segment, began offering the same set of supplementary services, such as room service, meeting and conference rooms, a swimming pool, a gift shop, personal grooming salon service, business services, cable, or Internet service in each room, these became commodity and part of the core product.

By March 2001, 78 precent of all Ritz-Carlton hotels offered highspeed Internet. While only two hotel chains — Omni and Westin hotels — have laptops available for guests, most hotels in that class offer two phone lines. Thus, over time, as competing firms add supplementary services, these previously differentiating value components become a commodity and lose their differentiating ability.

Consider the recent "advances" in the form of supplementary services in the hotel industry. Electronic kiosks can not only check you in with your room key and print your bill at checkout, but can also be an electronic concierge providing a guest with directions or maps to locations of interest. Human butlers are back in the hotel industry.

A growing number of hotels at the top end will provide a bath butler to prepare your luxury bath, a technology butler to solve your gadget problems, a private butler to pack your clothes or make your plans. If you are traveling with a baby or young child, you can find a hotel that offers a room for a nanny and kiddie perks such as baby-sitting, some one to read a bedtime story with live actors, or toys and activities for the child.

If you are bringing a pet along, some hotels will even offer you special pet services for a fee! Over time, hotels in the premium category will all be offering some or all of these supplementary features. These features become part of the standard offering and reduce themselves into a commodity for that category of hotels. You can expect that the "technology butler" service will be adopted by hotels for the business traveller segment, briefly providing a differentiating feature only very quickly to become standard fare. In designing superior customer value, a firm must translate customer needs to the specifications of a solution in terms of core and supplementary benefits.

All hotels need not have the same core product. That statement might seem confusing. Consider the Station Inn, in Pennsylvania, between Pittsburgh, and Harrisburg. It sits 125 feet from the railroad tracks. Railroading buffs sleep at the hotel just to see the freight trains, about 60 of them in a space of 24 hours. Same industry as hotels — but a somewhat different core product. Precise definition of a product can only emerge from deep inspection of the "What business are we in?" analysis.

The Supplementary Product

If product differentiation is derived from the enhancement that the supplementary product provides the core benefit, what constitutes supplementary product? Starting with the familiar: customer service is a common supplementary product (sometimes mistaken as the only element in the supplementary product). There are elements of the supplementary product, such as billing and payment, that are also common to all firms. They are required to facilitate the core product. These must also be viewed as part of the supplementary product to the extent that they affect customer value.

A firm can add to the customer value when its billing and payment process is customer focused or can detract from customer value if it is unpleasant in the customer's experience. Thus, there are some supplementary features that are found in all products that may not necessarily be recognized as such. Services marketing scholar Christopher Lovelock visualized this concept as a flower with petals. Think of the supplementary product as composed of elements of the total value bundle that facilitate and enhance (or detract from) the primary benefit sought in a product. Visualize these elements in three broad categories: facilitating services, quality of customer experience, and brand image.

Facilitating Services

Facilitating services are all those features of the product and activities of the firm that serve to facilitate the consumption of the core product. All firms have to provide some of these services at some level. Large retailers such as Sears, Walmart, and Lowe's will provide assembly and installation services for a fee and sometimes for no charge. The four broad categories of facilitating services — complementary services, customer education, customer access, and customer service. Customer-focused firms differentiate themselves by excelling in these facilitating services.

Complementary services are those related to the consumption of the core product. Some are almost necessary, like the waiting area in a physician's clinic or parking facilities at a hotel. Passengers on long international flights enjoy a wide range of inflight entertainment activities such as films, TV channels, and even casino-style gambling or electronic shopping and services such as being able to rent a car or book hotel rooms from the technology at their seats. Some complementary services are product differentiators.

For example, recognizing that some of its customers were coming in to the bank to conduct their business bringing little children with them, some Wells Fargo branches have a play area with television cartoons, Nintendo, and the like — and even sells toys to children. Washington Mutual offers calculators, pens, and piggy banks for a small fee.

To compete for advertising within its *Mutual Funds* magazine for instance, the publisher, Time, Inc. offers its advertising customers anything from primary research and use of its subscriber list to assistance in customizing gifts and promotions to readers or sponsorship featured on special cover-wraps. These are examples of complementary services that help differentiate value created by different providers.

Only with an intimate understanding of the customer's value chain consumption (and creation) activities can firms begin to see what opportunities there might be to facilitate the consumption of the core product. Sometimes complementary products are outsourced and provided directly to the customer by the supplier, as Wells Fargo did when it invited Starbuck's to open locations at some of its branches.

Ultimately, these complementary products ought to be consistent with the positioning strategy for the product. When a complementary feature is offered by all competitors in a particular product category, it becomes a commodity and part of the core product for that category and ceases to be a product differentiator.

Customer education is a form of facilitating service in that it provides information to the customer about the firm or the product in such a way that it facilitates the acquisition and consumption of the product. Instruction manuals and product support via telephone, fax, or the Internet are examples of customer education. Consider what some banks are doing to be more accessible for customers looking for assistance.

Tellers and managers are dressed in khakis and casual shirts to appear approachable and friendly. Bank of America has tested concierges in their lobby to help direct customer questions. Home Depot provides free classes on home improvement projects. Once again, the service feature might be a necessary component of the total product, but how it is designed and delivered may be a source of product differentiation.

Customer access has to do with everything that the firm does to make its products available to the customer and to facilitate the acquisition and consumption of the product. The hours of operation and location are a simple example of customer access for a service. The various touch points that are available to the customer to access the services of the firm or to reach someone within the firm would be a measure of the customer access. The methods of billing and payment for the product would relate to access to the product and can be a product differentiator.

For example, the convenience of payment by credit card was once a source of competitive advantage. Convenience and ease of use is a critical component of customer value for any product. Services such as financing options are also related to this idea of customer access. If these are not customer-focused, they will fail the customer and the firm.

Customer service is a key facilitating service and includes the commonly understood activity of businesses related to product complaints or 'failures, product returns, refunds, and such. The accommodation of special requests or adapting to unusual or irregular customer situations would be an example of customer service as well. Poor customer service reduces customer value and risks losing the customer.

Quality of Customer Experience

When the total product provided by all firms contains the same set of facilitating services, the product is not necessarily a commodity. The quality of the service offered remains a source of differentiation and therefore a supplementary feature of the product. Perceived quality enhances, and lack thereof reduces, the value of the core product. For example, even when Internet service is offered by all hotels, the quality of the customer experience with the Internet service is still a differentiator.

One hotel may require you to get into a closet, fish out the cables and force you to shape yourself into a yoga pose to hook up those cables to your laptop, while another hotel may offer wireless Internet access from anywhere on the property. All airlines provide seating and, depending on distance and class, also provide meals and entertainment while transporting you from point A to point B. Differentiation in seating can come in legroom and comfort. First-class services can offer six and a half feet of seat length, single seats, pajamas, and privacy partitions between seats. The quality of the entire customer experience is most definitely a source of differentiation.

Brand Image

Brand image is a supplementary feature of the product in that it adds to customer value. Brand image is the sum total of all the perceptions and attitudes about a brand. The image of the firm in the general media as well as for each individual customer provides a measure of the perceived quality of the product. The reputation of a firm is usually a result of the firm's product and actions.

Where there is very little tangible evidence of the product, as in the case of Web-based services, an entire industry of so-called "reputation managers" has emerged. These reputation managers are Web sites that rate the reputation of others! A firm's image is an implicit source of differentiation in the market. Brand image is the ultimate differentiator. When all else can be seen as equal, the brand image captures the essence of the difference in customer value among the alternatives available to the customer.

Firms have to determine what their total product offering is and what it should be. The value bundle of core and supplementary product needs to be designed based on what the customers expect as standard from all providers of a solution in a product category. The supplementary product might also contain features that are sources of differentiation among solution providers reflecting the positioning strategy of the firm. Firms must constantly watch the various solutions that are being offered. Sometimes the threat of competition comes from newer business models.

For example, online broker ETrade surprised the banking business when it began to open ATMs with new services. A deep and broad analysis of customer needs could reveal value-creating opportunities.

Matching Product to Customer Value Chain

To determine how best to match the total product as the superior solution to fit customer need, the firm must understand the customer's value chain. This is an integral part of the product concept development of the customer-focused firm. A thorough understanding of the customer needs and the customer value chain will help the firm conceive its product from the customer's point of view. Firms that best relate their own value chain to the buyer's value chain, said Michael Porter, will enjoy a sustainable differentiation strategy.

The value chain is essentially a chain of value-creating and value-consuming activities, where the value created as output from one activity becomes the input to another value-creating activity that in turn creates value as input for another activity, and so on. Thus, any activity can be assessed by the value it creates versus the value it consumes. This is why activity-based costing practices make a lot of sense. As, the value creating activities *within a firm* are those that contribute to the creation of either the core product or the supplementary product.

The value created by these activities is derived by the processing of the productive factors of the firm—its people, facilities, and equipment. There may be assets or value components sourced from external suppliers or intermediaries that contribute toward producing either the core or the supplementary product. Some outsourced services may even be delivered directly to the customer.

When any value component of the total product is outsourced, as in the case of a retailer offering customers outsourced financing options for purchases, or airlines outsourcing catering services, there is the obvious issue of quality assurance in the value creation that is outside the control of the firm. Externally (to the firm) when you relate other firms' value-creation activities into a value chain, you see that the value created by one entity contributes to the value created by the next entity in the value chain.

The same idea is referred to by economists, in the context of forecasting, as "derived demand." The demand for a product is dependent on the market for another product, such as aluminum and aircraft sales for example. A homeowner who gets a home improvement job done receives (customer) value from the home contractor who in turn is receiving value from the retailer who in turn receives value in the form of products and services from the supplier or manufacturer.

CUSTOMER VALUE OPPORTUNITIES

An understanding of the role that the firm's product plays in the customer's value chain could open up value-creating opportunities. A customer-focused firm has a detailed picture of the customer's consumption domain. When a firm views value consumption activities, it can see if there are other value creating opportunities that it can leverage from its assets. The watchmaker Swatch, for example, offers wrist watches with the technology that allows them to function as electronic passes at ski resorts in Switzerland or for public transportation in Finland.

Wrist gadgets can now serve not only the function of telling time, but also making phone calls, playing music and videos, browsing the Internet, or sending email. This example defies categorization of product. What is the product? It is a wallet or a pocketbook, as well as a telephone, a personal stereo, a personal VCR, and an Internet communications device! The exercise of determining what revenue opportunities Swatch's value-creating assets would offer forces a complete redefinition of the product.

The (product) solution is conceived to take advantage of opportunities to meet the needs in the customer's life with the assets that the firm has. Swatch has found a way to provide value to a recreational activity — skiing — and a functional activity — public transportation. It is contributing to the "customer access" components of the ski resort's and public transportation's product. By enabling customer access, information technology provides a whole range of supplementary benefits to a variety of products.

Contrast this with the example of a firm that does not understand the value of complementary product components and the customer's value chain. A customer came out of a movie theater in Kendall Square, Cambridge, and experienced a 40-minute ordeal trying to leave the theater's parking lot. She spent 30 minutes standing in line in frigid weather to pay the $2.50 parking fee and a further 15 minutes to exit the parking lot. When complaining to the management of the movie theater, she asked if she could pay the parking fee as she bought the movie ticket. She was told that the parking lot was owned by a different company and the theater would not take responsibility for the customer's bad experience at the parking lot.

In contrast, the airline SAS has been known to provide an annual dinner to taxicab drivers in Stockholm because the SAS management wants the drivers to treat passengers on the way to and from the airport with professional courtesy and respect. Clearly, firms that are customer-focused conceive their product differently from other firms, because of their intimate knowledge of the customer's consumption activities.

Information about the customer's consumption cycle, therefore, is a key prerequisite to exploring the opportunities that might be tapped. An understanding of how the customer actually benefits from the solution and the information of how, when, where, and with whom the customer consumes the product should provide some interesting revelations of what the firm is doing and can be doing in the composition of the total product solution for customers.

Thus, Procter & Gamble sends its researchers to homes to observe how people actually use laundry detergent. When Samsung was trying to break into the microwave business in the late 1970s and early 1980s, their design engineers observed homemakers shopping for microwaves at the retail store. You can stay ahead of the curve by offering value through supplementary product that your customer information tells you customers will be willing to pay for.

Even if the value-producing feature or activity cannot be priced separately, you may be able to command a premium for your superior customer value. Being customer-focused is critical in identifying opportunities for establishing superiority in customer value.

Conceiving and Designing

All decisions about the product and, therefore, the value-creating activities of the firm are based on an understanding of the value-consumption activities of its customers. What are the decisions and what are the issues to be considered in determining the total product by a firm? The three major decisions constituting the product strategy are: the product concept, the operations design, and the value creation and delivery process. The *product concept* defines the customer to be served and what value is to be provided. The *operations design* defines the productive assets of the firm required to create and deliver that value for that customer.

Together, the product concept and the operations design define the scope and configuration of the productive assets that can be leveraged to produce the specific customer value that maximizes profits to the firm. The *value creation and delivery process* executes the product concept with the operations design. In developing the product strategy, the firm makes a fundamental decision in answering the question of how the product will be positioned among all potential solutions to the customer's need.

This positioning question poses an asset- and market-based decision that comprises two perspectives.

* The *market-based perspective* looks outward at the market and asks what customer needs can be most profitably served by the firm.

* The *asset-based perspective* looks inward, at the firm's assets, and asks what assets of the firm can be most profitably leveraged by the firm.

The market-based perspective drives the product concept and the asset-based perspective drives the operations design. Thus, the initial step in developing product strategy involves two critical analyses. The product concept requires analyses of the various segments in the marketspace, while the operations design requires analyses of the firm's productive factors. These two sets of analyses are essential to determining what value-creating activities the firm should engage in to generate the maximum revenue from its assets.

The Market-Based Perspective

A market analysis to determine what would be the most profitable segment mix for a configuration of the productive assets of the firm is the foundation for product strategy. First you need to identify the segments and the solutions that are currently available to the segments. The target market selection or the selected market segments to be served can be based on the profitability and size of each segment that the firm's assets are best positioned to serve.

Here, a formal comparison of all the current solutions from the customer's perspective is necessary. Based on this analysis, the firm is able to determine what it can provide better than the alternative available to the appropriate target market, thus framing the firm's competitive advantage. Only after such customer needs analysis is it possible to specify what would be the desired customer value in the product offering.

The intended customer value in the product offering can now be translated into a detailed picture of the product concept—what the core and supplementary product ought to be.

Remember that the customer value also reflects the customer's implicit assessment of the firm's solution compared to competitive offerings and, indeed, all solutions that are available to customers in the segment. To ensure that the product concept can effectively be superior customer value, the core product should combine the imperatives that have become commodities in the product category with the appropriate features in the supplementary product to reflect superiority in customer value. Thus, the product concept embodies the product differentiation and the superiority in customer value.

The Asset-Based Perspective

The operations design decision rests on how best the assets of the firm can be most profitably leveraged and follows a sequence of questions that pertain to the productive assets and capabilities of the firm and how they should be deployed: What people, facilities, equipment assets are needed to create and deliver the product? Which employees' skills and knowledge would be needed? When and where would specific human capital be needed and for how much time? Similar questions are asked about the facilities and equipment of the firm. Of course, when the asset or resource is not available within the firm, it seeks suppliers or outsources that part of the value creation. Ultimately, the question is what competitive advantage the firm is capable of and how the assets needed should be configured.

Value Creation and Delivery Process

The process decisions are about determining the specific activities of the value creation and delivery. The process of making the product and delivering it to the customer must be detailed. The value creation and delivery activities required are set in a specific sequence. The firm must deliberate on the structure, content, and process of creating and delivering the product to the customer.

No product differentiation from supplementary product can be seen in isolation. If the total benefits from the product are not worth the costs that the customer incurs in acquiring and using the product, the product is likely to fail. The more carefully the firm designs the process with the customer in mind, the more likely the process ensures ease, convenience and quality for the customer. Thus, the customer-focused firm designs the creation and delivery process with the customer's perception of benefits and costs in the value consumption process.

Ensuring Customer-focused Value Creation

Firms must also recognize that since customer value is dynamic, they need to continually monitor and improve this customer value. Ignorance of the need to innovate to sustain competitive advantage is a common mistake committed by the complacent firm. The argument is sometimes made that the firm's priority of customer focus minimizes the attention to innovation. Customer focus and innovation are not contradictory, an either/or strategic decision. Indeed, to be customer-focused would mean that the firm is continually looking for new ways and solutions to meet customer needs — and to be aware that customer needs evolve as well.

Visualize what the firm needs to do to ensure that it is creating and delivering customer-focused value that can be sustained. To sustain superiority in customer value the firm must ensure that management continually assess its market and its assets to ensure that the choice of customer and the value being created and delivered by the firm maximizes the profit goals of the firm.

How well is the customer value in the product concept translated to the operations design and the value creation and delivery process? Is the superiority in customer value being executed. The answer lies in the customer's judgment. Customer-focused firms will let the customer decide whether the firm's value creation and delivery is customer-focused.

Continual customer satisfaction assessment information needs to be available for the customer-focused firm to improve by changing the product concept, the operations design, and the value creation and delivery process, or to continually reassess whether its assets and capabilities are being leveraged to realize the maximum profit potential. To ensure customer-focused value creation and delivery, firms must assess customers' perceptions of the benefits they receive as well as the costs incurred by them.

As an ongoing assessment of customer value, firms must continually assess customer needs and customer satisfaction. As customer needs change, the value bundle needs to be reviewed in terms of its product concept, operations design and delivery. An assessment of customer satisfaction presents an opportunity to improve customer value. The smart firm will continually monitor customer satisfaction to understand what customers perceive as the benefits they are getting from product compared to the costs that they incur.

Based on customer satisfaction, does the value bundle need to be modified and redesigned? Are the assets and capabilities of the firm being leveraged for maximum sustainable profits? If the expected customer value cannot be delivered with the existing value-creating assets, then the firm has two choices. Either it acquires or outsources the required value-creating assets and capabilities, or it determines that the target market decision needs revisiting. These questions are posed as a frame for a customer-focused analysis of the value creation and delivery of the firm.

Managing Customer Interactions

The process of delivering value is a tricky and detail-rich exercise; it requires careful planning, using techniques such as blueprinting to visualize the entire customer experience. From the customer perspective, some service encounters are critical incidents requiring more attention than others. All service encounters must be staged for a customer-focused experience just as in the production of theater.

Why does Kimberly-Clark manage the discount retailer Costco's inventory of its diapers? The firm has a salesperson live near the Costco headquarters, and a data analyst responsible for overseeing stock at 155 Costco stores in the western United States. Similarly, Procter & Gamble stations 250 people near Wal-mart's headquarters in Bentonville, Arkansas. Large retailers are asking suppliers to more actively manage the movement of products from factory to retail store shelves. P&G estimates that stock-outs amount to 11 percent of an average retailer's annual sales.

When firms like Kimberly-Clark pay more attention to how its immediate customers—the retailers—create value to their customers, they demonstrate that they are being customer focused. In fact, the Kimberly-Clark salesperson passed on information on how customers place packages in their shopping cart that played into package design for diapers. Wayne Sanders, chairman and CEO of Kimberly-Clark, attributes this change to the information age.

Prior to the industrial revolution, a service orientation and the individual-to-individual interaction was the predominant mode of competitiveness. Assembly-line production distanced the firm from the customer due to the sheer number of customers and the physical distance between the customers and the firm brought about by the wonders of modern transportation. Manufacturing goods became the engine of individual and collective (national) economic growth.

Mechanization far outpaced services and replaced producer-customer interaction, relying instead on intermediary institutions to provide a specialized set of competencies that the producing firm lacked. Businesses lost sight of the customer. Now, information technology has brought this full circle, back to the customer.

We call it the information age because what has changed in our time is that new technology has revolutionized the way information is handled. Another equally significant revolution is the change in the customer's domain. Customers not only have access to more information on products and service offerings, they also have the ability to interact with the providers of products and services in ways not previously possible. Providers can also present enhancements to customer experiences from the functionalities presented by technology. Service providers must, however, also manage customer participation when they utilize technology.

When firms take advantage of online marketplaces, it might be necessary to make changes in organizational structure so that value creation and delivery processes are adjusted to the addition of the online delivery. Two key technologies underlie this information age technology phenomenon: the Internet and wireless communications. What customer access to these technologies has done is to bring customer interactions to a new level and to the front and centre in how a firm deals with the customer. These interactions are essentially service encounters with the customer.

Much has been said about the service encounter — customer interactions with the firm. In a way, all customer relationship management (CRM) solutions are basically technology support to ensure that the firm maximizes returns from customers by enabling customer-focused interactions. The service encounter is truly "where the rubber hits the road" — where the prospects of customer loyalty are materialized or lost. It is where promises made in advertising are honored or reneged on, and where expectations of customers are disappointed or met. It is where all the assets of the firm need to be brought to bear.

Service encounters with the customers are also laden with the challenges of a product that is produced and consumed in real time, where failures are bound to happen. When the firm designs the value creation process with the customer in mind, it will be prepared for all predictable eventualities. When the service fails, smart firms have smart processes, that recover and learn from the failure. They have service recovery and knowledge management processes in place.

As firms compete more and more on services, the management of the customer interaction becomes critical to ensure superiority in customer value. As the core product is a commodity, and most facilitating services approach the commodity state, the competitiveness comes from how the customer is treated by the firm at each and every encounter.

This discusses a method for designing the value delivery process with special attention paid to the service encounter and the critical incidents in the value creation and delivery process. A framework for designing the service encounter based on the theatrical metaphor is offered as a way to examine the customer focus of value creation and delivery.

The Ingredients of Customer Interactions

Analyzing customer interactions can be very rewarding. Amazon constantly tracks the reasons for every customer contact. It has made numerous changes to its value creation and delivery process based on customer input. The customer service function at Amazon is considered a research lab for ways to improve the Amazon customer experience. We can turn to the unique characteristics of services to obtain a clear understanding of what is involved in a customer interaction. Something intangible is always exchanged in the interaction (intangibility). Without a customer, there is no interaction and the separation of production and consumption is irrelevant (inseparability).

To conduct the interaction, the firm has the resources and infrastructure in place that "perish" when not utilized (perishability). Each interaction is unique (variability). From these characteristics, emerge situations that are challenges and opportunities typical to the service firm. We need to keep these in mind as we design the process of creating and delivering customer value. At the most general level, all customer interactions occur in a certain place (or space), at a certain time, for a certain duration, between certain entities, in a certain manner for a certain purpose.

Consider these as ingredients in a customer interaction and categorize them into elements of structure, content, and process. When the customer interaction is mapped onto the consumption activity cycle, the objective of the interaction becomes the driving force behind the design of the interaction in terms of its structure, content, and process.

- *Structure* relates to who or what entities are involved in the interaction.
- *Content* relates to the subject, the task, and its significance in the interaction.
- *Process* relates to the sequence of steps in the interaction.

The primary decisions that would configure the structure of the interaction involve the nature of the interaction—what entity should represent the firm, should the interaction be face-to-face or not, what technology can be utilized, where should it occur, and so forth. As a complementary question about the customer: What entities from the customer's domain are involved in the interaction? Similarly, decisions determining the content of the interaction should centre around the task and include what information should be required in the interaction, what should transpire between the parties, and what value should be created and consumed in the interaction.

Decisions about process would include a script of the interaction and the roles of both parties in the interaction. Any of these decisions must be based on the information that the firm has about the customer's consumption cycle and on preferences regarding these dimensions of the interaction. For example, technology may provide the firm with scale economies, but the customer may want to disengage from the script and interact with a person instead. Ultimately, the important question is what value is being created in the interaction.

THE VALUE CREATION AND DELIVERY PROCESS

As, the execution of the product concept as the delivery of the product involves a process, a sequence of activities. Therefore, designing the product must necessarily include a design of the delivery process. Lynn Shostack proposed a technique called "blueprinting" to detail the process design in the delivery of services. The service blueprint is a map or flowchart as a visual representation of the process of service delivery.

Essentially, the blueprint design is diagramed in three steps. In the first step, all types of customer interactions with the firm are listed. Next, these interactions are arranged in sequence of occurrence. In the final step, processes within the firm that are required to create and provide each customer interaction are mapped. How and what information, materials, facilities and equipment, and people are processed must be included in the service blueprint.

The customer-focused blueprint is one that is designed with the customer in mind. With the activities in the customer's consumption cycle in mind, processes within the firm need to be designed in such a way that the desired value added in the solution is reflected at each and every encounter with the customer. Commerce Bancorp, for example, encourages its employees to make suggestions in streamlining the service delivery process. Commerce Bancorp employees receive a fifty-dollar reward for finding something about a process that does not contribute to customer value and instead is an impediment to serving the customer; the bank immediately adjusts its operations.

Even as a simple representation of the value creation and delivery process, the service blueprint has a variety of important strategic, analytical, and diagnostic uses. In designing the delivery of the product, it can be used to determine the uniqueness of the service process. When compared with competitive blueprints, where and at which points in the process the firm derives its competitive advantage and superiority in customer value become evident.

Thus, the service blueprint can be used in a strategic analysis of the differentiating features and benefits of the service delivered by the firm. Only with such comparative analysis of the service blueprint is it possible to ensure that the value creation and delivery processes are commensurate with the superior customer value planned in the product concept and operations design.

The service blueprint has two major dimensions that characterize the nature and scope of the service delivery process: complexity and divergence. The complexity dimension reflects the number of steps or activities in the process. A highly complex service process has a great variety of activities during the series of customer interactions. Some of these activities occur in the front stage and some in the back stage, away from the customer's view.

A so-called line of visibility determines what the customer sees or does not see. The customer-focused firm will need to examine everything that the customer sees and experiences to determine whether the service delivery at the service encounter is contributing to or detracting from customer value.

The divergence dimension reflects the degree of flexibility at each customer interaction. The greater the divergence, the greater the customization built into the service—the greater the number of options the customer is given to choose from at each interaction. Therefore, supplementary product benefits such as customization must be mapped into the service delivery process.

Complexity and divergence decisions will determine the position of the firm's offering compared to competitive offerings. Competitive blueprints can be compared to determine how competitive value bundles differ from each other. It becomes clear that the positioning of each competitive offering is reflected in the service blueprints.

The patient experience is very carefully orchestrated to support the entire concept of the "Shouldice Method." The surgical procedure itself is probably the least sophisticated element of the whole process. The recovery process is more important than the surgery. Every step in the blueprint is designed for a speedy recovery and an overall "enjoyable" experience—despite the anxiety and pain of the surgery. The method is known for a very low recurrence rate and high customer satisfaction. The hospital even has patient reunion parties each year that are sometimes attended by as many as a thousand former patients.

For this service, its critical steps are not in the surgical procedure but in in the recovery process and the total customer experience. All customer interactions depicted in the blueprint are not of equal importance in terms of criticality toward customer value—some are more important to the customer than others. It is imperative that the firm employ its knowledge of customer expectations so that the critical interactions are highlighted for priority when allocating resources of the firm.

This is the primary application of the blueprint technique. Jan Carlson, CEO of the SAS airline, popularized the notion of the "moment of truth" in services.

While each customer interaction is a moment of truth, those product features and customer encounters with the firm that are considered critical to the satisfaction of the customer are considered *critical incidents*. The critical incident technique has been applied in the managing of services to understand what can make or break the customer's experience. The method was initially developed for the U.S. Air Force in simulations to help in cockpit design and pilot training.

A simple question to the customer soliciting a verbal description ofmct is stocked and well displayed and that the floor salespeople are presenting the product to the customer in the way Handspring had instructed them to do. Microsoft hands out a free Casio pocketPC to every sales rep who completes its training. These examples show an awareness that customer interactions at these intermediaries are failpoints and need to be monitored.

Critical incidents help in assigning the appropriate level of resources within the firm. Thus, the service blueprint is useful from a quality control perspective. Customer evaluations of a firm's service can be linked to specific customer encounters in the blueprints and all processes directly affecting or affected by the critical incidents must be monitored for quality. Customer-focused quality control efforts can be focused on critical variables that can be controlled. Service recovery procedures must be designed into the process at these points.

Interactive technologies provided by the information age revolution are now an integral part of the design of the value creation and delivery process. These technologies work through the three primary productive components of customer value — the people, the infrastructure in facilities and equipment, and processes and systems of value creation and delivery. Consider all the ways in which the customer can interact with a firm. Each touch-point can be made by the converging technologies of today. For example, in some interactions the location of either buyer or seller becomes irrelevant.

Customers can interact by wireless or wired means, which can be by voice, video, or text, initiated by either party; they are constantly getting cheaper and better in features and functionality. The basic difficulty with any of this newer interactive technology is that the objective and role of the technology can be lost sight of and the technology itself become the focus. The final measure is whether the technology is contributing toward sustainable profits for the firm by improving customer value.

The benefits of interactive technology come in the form of efficiency and efficacy. Efficiency is reflected in increasing productivity and efficacy is reflected in improved customer value. Both of these outcomes contribute to the bottom line of the firm. For a firm competing on service, interactive technology is critical to sustain profits because of what it can do to customer value.

Maximize the return from interactive technologies by deploying it at the critical points; prioritized by its customer value-creating power. For example, a common source of dissatisfaction and irritation to the customer is the length of waiting time before being served. Any point in the process where there is likely to be a wait is a potential critical incident requiring the firm's express attention. If there are interactive technologies that can be used to effectively manage the customer in these situations, customer value is not threatened.

The Problem of Waiting Time

All the service components in the total product, being susceptible to the inherent characteristic of being produced and consumed in real time, force customers to the experience of "wait." As, demand and supply need to be matched as closely as possible. To avoid idle capacity, services adopt the queuing (and scheduling) approach to processing customers, resulting in a certain amount of wait before the process begins. In multi-stage processes, which many services are, in-process waits are common as well. The problem of course is that no one wants to wait.

Customers will wait for a certain amount of time that is reasonable to them as per their expectations. What is reasonable is a very subjective judgment, of course—one that requires analysis of the factors that determine the perception of waits.

Maister's work on the psychological cause and effects of wait suggests a way to reduce the negative consequences where wait is a given.

- Preprocess waits feel longer than in-process waits. Allow the customer to begin the process by completing the first step in the value delivery process.

- Unoccupied waits feel longer. Keep the customer in wait occupied with activities, preferably related to the product or service.

- Uncertain waits feel longer. Keep the customer informed about how long the expected wait is to likely to be.

- Unexplained waits feel longer. Keep the customer informed about why they are having to wait.

- Inequitable waits feel longer. Serve customer based on priority, determined by what is considered just in the ambient society.

- Value of the service. Ensure that the customer value you are providing is clearly superior to any alternative.

The most important common thread through all of these suggestions is that they all have to do with managing the customer perceptions. The firm is better placed if the customer perceptions can be influenced favourably by managing customer role and expectations in the value creation and delivery.

The Customer-focused Firm

To varying degrees, customers play a role in the production of products and not just in the consumption. Firms that realize this have paid serious attention to how the customer fits in the organization. In services, the role of the customer is abundantly evident. Customers engage in a coproduction role. In services or in the service component of the product, customers are coproducers in self-service configurations.

All services as we saw earlier involve some kind of customer participation, as in the case of providing required information to the tax preparer, for example. Even when placing the order with the provider, the customer is essentially contributing to the production of the product by providing input into product specifications.

When the customer is interacting with the firm and this interaction has some relevance to the production of the product, by definition we have the customer engaging in coproducing the product. Service firms instinctively manage the role of the customers in their participation in the service.

Restaurants have menus, professors have syllabi, airlines have their method of emplaning passengers according to a certain priority, etc. Sometimes managing customer roles may be in the form of a complementary service. For example, Universal is experimenting with a calming zone adjunct to its roller-coaster rides, for people who need help with coaster-phobia before they get on the ride.

UPS and other package delivery firms instinctively accommodate special customer requests as to where their packages should be deposited if the customer is not at the location of delivery. It is a matter of perspective. If we take the notion of value creation a step further, we could argue that even in the case of packaged goods, the value hasn't really been created to the customer until the customer actually begins to use the product.

Value creation and value consumption are both inextricably interwoven into a continuous and overlapping set of activities. This is an important issue in a service orientation. In the case of services, customer interaction is a given. And the firm must include the customer in the design of its operations.

The customer-focused firm by always keeping the customer in central view manages the role of the customer in the value creation and the value consumption activities. For example, providing excellent product assembly instructions improves the value created for the customer in an "easy to assemble yourself" product.

Conversely, even with a well-engineered product, poor instructions reduce customer value. Recent product innovation studies have shown that some firms are equipping customers with the tools to design and develop their own products, ranging from minor modifications to major new innovations. PTC (formerly Parametric Technology) is a world leader in computer-assisted-design (CAD) technology, whose software solutions bring suppliers and customers together so that a client can use the software during product development to virtually interface with upstream and downstream players.

Ultimately, anything that the firm can do to manage the customer's role in the value creation as well as value consumption process is an imperative for the customer-focused firm. As mentioned earlier, Procter & Gamble is recruiting families to allow a team of their ethnographer-filmmakers to observe their daily routines to get a better sense for how their customers use household products.

The actual use situation says a lot about how a product should be designed. It would also be a way to learn how the customers should be instructed in the appropriate use of the product. How does a firm take a systematic approach to managing the customer? First, the firm identifies all the activities that the customer enacts in the value creation as well as value consumption blueprint.

Next, the firm scripts the role for the customer at each activity. Firms then have to determine how best to communicate and educate the customer in playing the appropriate role. Customers' performance of their roles will depend on their ability and inclination to perform those roles. The consumer's ability will depend on consumption skills that the customer's past experience may have provided.

The customer's inclination will depend on the personal characteristics of the customer. Familiarity with the service and service provider could affect the ability and the inclination of the customer in the role performance. In the case of services and the service components of a product, customer role is managed by the service provider. The service provider's production skills and motivation could affect the way in which the service provider manages the customer's role.

The service provider's familiarity with the segment and the specific customer will influence the service provider's skills and motivation in managing the customer's role performance. The service provider's production skills are reflected in its operations and customer interaction skills. The personal characteristics of the service provider can affect the service provider's motivation. When designing the service delivery process, it is important that the firm incorporate these factors on the customer's role performance.

Staging the Customer Interaction

The essence of customer-focused design of the value delivery process is in the design of the customer interactions that make up the blueprint. Customer interactions have gone by various names in research on services – service encounters and moments of truth, for example. Designing the customer interaction requires analysis and decisions regarding three of the structural dimensions of the value delivery process: the process itself, the participants, and the physical facilities and equipment. Customer roles have to be defined so that customer participation can be designed into the delivery process as appropriate.

One useful framework that has utility in managing customer interactions is to view service encounters metaphorically as "drama." When firms and customers interact, firms try to put on the best performance – like theater. The firm's *actors* perform their roles for the *audience*, the customer. In any service, there are actors and audiences involved in producing *performances* in a *setting*.

For example, restaurant customers are audiences with hosts and waitstaff as actors in the physical setting of the restaurant. Just as in theatrical production, there is a stage (the dining area) and a backstage (the kitchen); there are acts (welcome and seating, appetizers, main course, and dessert), scripts (menu and specials), props (menu card and table setting), and costumes.

Each element is carefully choreographed to produce the intended impression on the audience. Any service provider can examine each of these elements and determine the exact impression that would be appropriate for each customer segment. The "drive-thru" or take-out type of fast-food restaurant confronts different impression management challenges compared to the sit-down type of restaurant.

Different types of services would require a different combination of these theatrical elements for the desired impression. Services rendered at arm's length over the telephone or by mail, such as in investment brokerage services or credit card services, would see these elements differently from the face-to-face setting of hotels or airlines.

Internet "clicks" retailers would find a different set of issues compared to the "bricks" retailers. Hybrid retailers have to approach the issues differently for the two different settings. In some cases, the performance is shared by a number of customers, as in a restaurant or a hotel or airline where customers share many aspects of the firm's productive factors—employees, facilities, and equipment.

The setting dictates a shared product, and some aspects of the setting are visible (onstage) to the customer and some are invisible (backstage). Performers or actors (frontline employees) utilize scripts, costumes, and props in their performances. Every feature or theatrical element is carefully choreographed for the intended impression.

Therefore, just as in theater, managers can focus on impression management for customer-focused business activities. World-class service firms like Disney employ a service orientation profile (personality and skills) in their hiring; they call their employees the cast and carefully train them in both operations and customer interaction skills.

Disney's "guests" are treated in a customer-focused manner evident in the culture, in the way the cast is empowered and motivated to provide customer-focused performances. What is the type of setting where the customer value is delivered? There could be no interaction necessary with the value provider at certain phases in the consumption cycle, and there could be phases of physical, face-to-face interaction, or electronics-mediated interactions. At each interaction, the value provider must determine the line of visibility between onstage and backstage, what is or should be visible to the customer.

What is the process and duration of the various steps in the interaction, and what kinds of orientation must there be for the participants? What is the customer role in each setting, and are there other customers in shared experience?

Ultimately, all of the theatrical elements need to be carefully directed and choreographed to provide the appropriate impression. *Servicescape* is a term that refers to the immediate physical and social environments containing a service experience, transaction, or event.

The servicescape plays a number of different roles such as packaging the value bundle, facilitating the value delivery activities, socializing the employee and the customer, and differentiating the product. Managers must understand which servicescape elements can be controlled for the desired behaviours from both employees and customers.

The process of delivering the product must be analyzed, using techniques such as blueprinting and choreographing impressions on the customer. What is the role of each process in the value creation and delivery as seen by a firm with a customer focus? What should the structure and content of the process include by way of design?

What are the critical points in the process for its meeting customer needs and preferences? The firm should have in place metrics and customer-determined benchmarks that will trigger recovery procedures, anticipating that the firm will sometimes fail the customer. A boutique hotel set in the heart of downtown will win the race for new and repeat business with the help of customer-relationship management. Beating the competition is not an easy task in the highly competitive world. The Hotel Customer service always has been king at the room property known for unique amenities.

Using the theatrical metaphor, any impression that the actor, the audience, the setting, or the performance imparts must be examined and managed by the firm. The focus is on the effect each customer interaction or service encounter has on customer value from the customer's point of view. Customer-focused firms design the value creation and delivery process with the customer in mind.

Chapter 5

Customer Loyalty

THE LEARNING CONCEPT OF THE HOTEL

The potential benefits of a branded customer experience to the organization are seen in the kinds of measures that directly influence revenues, profits and shareholder value, such as higher margins and increased share of spend. If a $1 billion enterprise increases its investment in customer interactions from average to high, it can anticipate a $42 million return on the investment, according to a study by Accenture and Montgomery Research.

Today, however, an increasing number of hotel owners and operators are discovering the value in promoting their environmentally friendly guest rooms as EcoRooms. What exactly is an EcoRoom? An EcoRoom is a guestroom that includes at least a dozen products that are energy efficient, water efficient, waste reducing, non-toxic and/or biodegradable.

Moreover, the P/E ratios of most companies with above average customer loyalty index scores are more than double that of their competitors. Some of W. Edwards Deming's 14 points counseling companywide continuous learning include: —Improve constantly and forever every process for planning production and service, —Institute training on the job, and —Institute a vigorous programme of education and self-improvement for everyone.

In 1990, Peter Senge put forth the case for continuous learning as a means for staff to perform their jobs better, solve problems, deal with process issues, face and counter threats, and capitalize on opportunities. As Karl Albrecht has said, "the tremendous diversification of work, and the fact that more and more jobs involve using knowledge and skill to create value rather then just following pre-programmemed tasks, means that managers must devote much more attention to the way people work, and, reluctantly in many cases, to the way they think and feel." In his 1993 book, *Post-Capitalist Society*, Peter Drucker said: "The basic economic resource—'the means of production' to use the economist's term—is no longer capital, nor natural resources (the economist's 'land') nor labour. It is and will be knowledge."

Corning, Inc., for one, believes in the value of knowledge. Between job and classroom, all employees must spend at least 5per cent of their worktime training. Tracked by management on a business unit basis, employees now average over 90 hours of training each year. At the heart of this customer experience is a brand promise that goes far beyond product or service attributes to a total relationship that creates an emotional connection.

The goal is to have marketing bring in customers with promises that the rest of the organization – most importantly those employees who interact with customers – can keep. I say "most importantly" because employees are the brand ambassadors who most directly influence customers' impressions. Consumers rate people as the most important determinant of customer loyalty to brands, according to a survey by The Gallup Organization. They also rate customer service as one of the most important factors influencing an excellent customer experience, exceeded only by actions a company takes in response to a problem or request, according to a survey by The Forum Corp.

Within the learning company, every employee is a knowledge worker. Every employee is involved with the business. Every employee is trained and motivated, can work in teams, can be flexible and innovative. Every employee supports the overriding goal of generating optimal customer loyalty. These are core competencies, the skills by which a company seeks opportunities and solves problems. This even extends to leaders who need high-involvement skills with customers and customer support processes so that they are not shielded from reality.

The company, as a process, acts to facilitate the learning and experience of each employee, and the employees, as a process, help to develop and facilitate change, adaptation, and even transformation. Peter Bonfield, CEO of British-based ICL, has used the term *resiliance* to describe the learning company. It is able to transfer intelligence throughout the organization, adjust direction as needed, and take advantage of opportunities presented. He concludes:

It is absolutely imperative that our people are able to respond and react in an innovative and entrepreneurial way. We have to recruit the best, nurture them and then let them create new opportunities for themselves and for ICL.

The more we focused on the customer, the more we realized that we had to become a fully open company. . . . Because we are open, we are able to embrace new concepts swiftly and efficiently. By developing our people to think in an open way they are much more flexible and responsive to change. They embrace change willingly so that they can meet the challenges which come with openness.

The company, as well as individual employees, builds its collective skill level by proactively embracing, creating and responding to change, and by what James Higgins terms *knowledge management*. "This means identifying knowledge resources, creating new knowledge, and disseminating knowledge throughout the organization. It also means taking the tacit knowledge of each employee about how to do his or her job and turning it into explicit knowledge that others may use."

In the learning organization, there is a strong interrelationship between individual, group, and company skills. Not only are skills created with enthusiasm and learning gathered from multiple sources, but skills and insight are transferred quickly and effectively throughout the company. Involved employees throughout the company develop a knowledge and breadth of understanding about processes and relationships. Learning becomes an everyday fact of life, part of the overall process of work. Employees are cross-trained, function in cells or flow lines — natural teams — and can easily train one another or trade job functions.

As discussed, where learning on behalf of customer loyalty is a focus, organizational boundaries blur. In most companies, systems analysts, engineers, secretaries — and on — and on — are stuck in their jobs, pigeonholed throughout their careers. Their potential benefit, and their benefit to customers, is stunted. In learning organizations, flexibility is featured. Sales people can move to customer service, customer service to credit, credit to sales, production to information systems. Motivated staff members get to pursue career interests, their jobs are kept interesting and challenging, and they are inspired to generate novel, useful ideas.

Companies interested in basic and advanced skills enhancement must provide formal training for everyone. The training needs to be continuous, since changes in technology, management theory, and work processes are ever-changing. State, federal, and municipal governments, plus colleges and universities, offer assistance programmes and training. Training is available from professional organizations offering a broad spectrum of instruction, from Total Quality and customer service to computer software and machine maintenance. Large companies — such as IBM, McDonald's, Milliken, General Electric, and Motorola — have their own training "institutes."

Rosenbluth International's training programme is called Learning Frontiers, which provides instruction on culture and service. The programme is so successful, it has become a line of business for Rosenbluth. "Training provides a more proficient work force, improves quality, and cements loyalty.

We attribute much of our success to our training Philosophy and programmes. We believe that our approach to learning magnifies the contributions of our people, makes our business more profitable, and helps us achieve our goals."

Some of this company—sponsored training, at least in the United States, is a response to the level of workforce competency created by the education system. A 1991 report by the federal government called SCANS (Secretary's Commission on Achieving Necessary Skills) concluded:

The message to us was universal: good jobs will increasingly depend on people who can put knowledge to work. What we found was disturbing: more than half our young people leave school without the knowledge or foundation required to find and hold a good job.

The SCANS report identified a three-part foundation set of skills and qualities and defined five areas of competence. While it could be argued that the report places too much responsibility on the education system, nonetheless it does reflect evolving needs of business. As noted by quality expert Philip B. Crosby: "Today, we are in a business world where perhaps 80 percent of current jobs did not exist fifteen years ago. They probably cannot be completely learned in school and are probably not taught there. . . . So the individuals have to keep going back to learning in order to retain mastery of their trade."

Companies must first have individuals who want to learn, and this is where human resources departments can be of real assistance. Their role in screening and hiring, support and management of training programmes, and even selection and consultation with individuals for specific training activities is very important. They also help assure that learning takes place within a strategic context, that is within and for the goals of the organization.

In addition to problem-solving skills, which can be tactical and limiting (and focused on deficiencies), learning should be appreciative. That is, it is an expressive inquiry that envisions possibilities, capitalizing on those things the company does well.

Group learning and collaborative learning (such as W. L. Gore's one-on-one mentoring programme), as well as training for individuals, takes on greater meaning as companies utilize greater networking, cross-functionality, and teamwork. In addition to conceptual and task training, individuals must have interpersonal and interactive skills. Task-oriented group learning enables individuals to acquire new capabilities while broadening group process exposure.

Tom Peters' alternative term and concept for the learning organization is "knowledge management structure," or KMS for short. To roughly paraphrase and interpret his approach, Peters sees the value of knowledge and learning as much, or more, from the relationships established as a product of the creation of employees' experience, information, and insight as from their expertise. This translates, in part, to more networks and teams and fewer bureaucracies. KMS creates and identifies internal experts; uses their knowledge; packages, systematizes, and distributes the information they and others create; and supports the network structure and culture required to keep it flowing through teamwork, training, and purposeful cross-group sharing. He summarizes the four key areas of learning process:

Systematic knowledge capture and dissemination— Some of this is related to the customer information system (or more broadly applied management information system). Does the company have a system and structure for obtaining, storing, and using information, knowledge, and insight? As Jan Carlzon, chairman of Scandinavian Airline Systems has said: "A person who hasn't got information cannot take responsibility. A person who's got information cannot escape from taking responsibility.

Learning with clients (customers)— Beyond the traditional and routine approaches to qualitative and quantitative research, does the company—like Weyerhauser,

Levi Strauss, or John Deere—create continuous learning and interchange opportunities with customers?

Learning from outsiders— Using the resources of the community (academic), consultants, and professional trainers.

Learning from each other— How well does the company create knowledge within and pass it from individual to individual, group to group, department to department, and division to division?

The formula, then, for companies seeking employee skills that are customer-driven and commitment-based is straightforward: train, develop, involve, and recognize.

SKILLS AND CAPABILITIES

Mobil Oil conducted market research among over 2,000 motorists and found that only 20per cent buy strictly on price. Those price shoppers spend only $700 annually at service stations; however, most buyers, while desirous of competitive pricing, wanted things like more human contact, quick service, and attendants who recognize them. After benchmarking service-oriented companies like Ritz-Carlton Hotels and Nordstrom department stores, Mobil introduced their new strategic concept, called Friendly Serve. Friendly Serve attendants are now at many Mobil stations (as many as 85per cent of Mobil's dealers will eventually be in the programme). They have been specially trained to be customer proactive— pumping gas, cleaning windshields, and getting coffee for customers.

The concept also includes better lighting and cleaner facilities; but, having service-oriented attendants, alone, has increased sales at many service stations by 15 to 20per cent.Companies achieve competitive advantage principally when their customers perceive a value in them higher than other companies. Achieving competitive advantage is the responsibility of the entire company and every individual in it. This means that, not only does the company have to be (at least) current on its knowledge of customer perceptions and be perceived as unique, it must also demonstrate capabilities and skills individually reflective of that uniqueness by staff members and by the overall organization.

Finally, and of greatest importance, these capabilities must be perceived by customers as creating value. In *Competing for the Future*, Gary Hamel and C. K. Prahalad identify core competencies as a company-wide bundling of skills and technologies. They have hypothesized that, to be a core competency, a skill must meet three tests.

Customer Value — if the customer does not derive specific benefit from a skill, it cannot be considered a competence. For example, Motorola's rapid and customized production cycle time is perceived by customers as value, so the collection of skills required to provide it are a competency.

Competitive Differentiation — L .L. Bean's and Nordstrom's service skills make them competitively unique when compared to other companies. Their customer service goes well beyond the basics, and they are always seeking new ways to improve. Mobil's advertising emphasizes the personal service available at its service stations, differentiating itself from competition.

Extendibility — Does the company have strategic flexibility, the skills necessary to move into new or related market opportunities? Several of the major auto manufacturers, or instances, have worked for years on battery powered cars so they would have a marketable product at the appropriate time. Other less flexible companies, notably in high tech industries, have gone the way of buggy whip manufacturers, unable to extend their competencies as mar kets or customer needs evolved. Apple, for example has receded to only a few areas of application advantage, while Microsoft continues its impressive growth.

These competencies very much the SOCAP/Maritz study finding first discussed in the Introduction. To generate optimum customer loyalty for the company, employee skills must be aligned with customer needs, problems, expectations, and complaints.

Competencies are transferred to strengthen positions with customers. They are acquired to protect or extend franchises in existing markets, as QVC Network has done to broaden its attractiveness to at-home shoppers and as MBNA has done to offer superior credit card service. Kodak has done it to compete in the digital photography business. Competencies, with equal fortitude, may need to be excised if they become obsolete or off-strategy for customer need alignment.

Organizational capability can be expressed in terms of strategy, economic or technical strength, structure, or leadership approaches; however, the most visible demonstration to customers is through staff interface and other areas of direct performance. Several skill sets are required: responsiveness, relationship building, management and human resource practices, flexibility (including the ability to learn flexibly — individuals, in teams or projects, cross-training, on-the-job or in classroom settings), and the capacity for change. This begins during the hiring process. Customer sensitive people are self-responsive, capable of independent thought and achievement, able to overcome setbacks, and non-blaming. They are people with goals and ambition, with positive self-esteem. To use a term by now familiar, they are empowered — or can easily adapt to an empowered culture.

Many service-based companies — financial services, healthcare, foodservice, lodging, car rental — have determined that the capability of their organizations has been defined by staff responsiveness. Customers infer that the company is or is not responsive, and thus worthy of loyalty, based on performance during transactions. Responsiveness is also based on the level of cooperation, communication, and support employees exhibit for each other. Responsiveness may be tangible (time, completeness) or intangible (feelings). By extension, it may be proactive as well as reactive.

Relationships are influenced by collective and individual skill levels. When customers have had long-term relationships with companies, they often come to depend on and expect certain skill levels and positive attitudes from their contacts. If that contact is lost — retirement, downsizing, firing, changing jobs, and so forth — it may be very difficult to sustain the image and loyalty. Managers must use the right tools, or levers, to influence inside and outside customers. The tools are used to *create* competencies within organizations by hiring and training. They *reinforce* competencies through evaluation, reward, and recognition. Finally, they *sustain* competencies through organizational design and methods of communication.

Change is coming at companies from many directions — environmental and regulatory influences, workforce availability and mindset, new approaches to organizational architecture — to identify just a few. The most significant changes, however, are those created by everchanging customer needs and requirements. Skill level is also judged by how well leaders and their companies anticipate and facilitate change. Saturn is an example of a company that has created the skills and capabilities necessary to generate customer-perceived value. Although General Motors invested huge sums of money to create cutting-edge engineering and manufacturing technologies, it was ultimately the direct performance skill sets that made the venture work.

First, General Motors worked closely with the auto unions to select employees with preexisting team and responsiveness competencies. While early car recalls showed some lingering hierarchical General Motors' culture weaknesses, the Saturn culture of training and customer focus has now been established. In *Win the Value Revolution*, Robert Tucker reviewed the skills development strategies of companies like Southwest Airlines, Levi Strauss, Home Depot, ServiceMaster, Rubbermaid, and Intel. They all believe in cultures that generate customer value through people. Diversity, recognition, high morals and ethics, and empowerment are common themes. Tucker identified four key attributes of such companies:

1. *Establish a continuous value improvement process:* "Undertaking a value improvement programme begins by instructing your employees — even those who've never come into contact with one of your

customers—to understand more about the overall business. Your team needs to know the impact of their work on the value perceptions of customers." At the Long Beach Medical Centre in California, new residents are admitted to the hospital under assumed names and with falsified symptoms. The idea is for them to experience, over a 24-hour period, hospital treatment from the perspective of patients. This creates a more patient-sensitive corps of residents.

2. Teach your employees how to "own their own employability" This includes identifying the added value that each employee, team, and department provides to the end customer, and also identifying the skill levels needed to provide that value.

3. Teach managers that serve internal customers how to add value. "Any department in a company that doesn't directly serve the external customer needs to reposition itself as a business within a business." This is particularly important for training and HRD managers, who must be proactive in helping staff develop change processes, diversity, productivity, and related skills—capabilities needed by an organization wishing to improve its level of value delivery to customers.

4. *Treat employees as customers:* New workforce issues— longer working hours, more single working parents, and the like—necessitate that firms look at employees as customers. As Rosenbluth International sees it, making employees more skillful and valuable equals more employee loyalty which equals more customer loyalty.

At companies like State Farm Insurance, for example, customer liaison and supervision staff may take on diversity assignments—such as learning about foreign cultures—and share the knowledge with associates. They might also, at the same time, take advanced computer skills training at the company's headquarters in Bloomington, Illinois, or via a live or video instruction programme. This type of learning achieves all four of Tucker's described attributes. It adds value for the employees and managers who, in turn, add value to internal and external customers.

Hewlett-Packard has taken much of its sales training out of the classroom and put it into the field, so that sales representatives can spend more time with customers. H-P frequently introduces new products and in the past it had to bring the sales force into conference centres to learn about them. This required up to three weeks a year, creating breaks in customer relationship continuity. Several years ago, the Hewlett-Packard Interactive Network (HPIN) was created, so that training was made available on line, to fit the sales representatives' schedules.

3M, following a practice, developed a totally customer-driven approach to individualized sales skills training. Called A.C.T. (for Assessment, Content, and Training), the process begins with debriefing customers on their perception of representatives' product knowledge, interpersonal skills, strategic capabilities, negotiating abilities, and teamwork. Completed customer questionnaires form a summary report for each salesperson, and the salesperson designs his or her own training programme according to the biggest gap between perceived and desired performance. The salesperson also reviews the (grouped) results with customers, thereby reinforcing the relationship.

Federal Express sales managers and sales representatives developed the Global Customer Learning Laboratory. Working with customers, they developed "Learning Partner" programmes for key accounts to develop new approaches for handling logistics issues.

Southwest Airlines' University for People recently introduced its "Mind the Gap" programme for the entire employee force of 22,000. "Mind the Gap" (the term comes from the London subway system, where announcements are made to passengers to be aware of the space between the platform and train when boarding) focuses on perceptual differences between Southwest employees and passengers, and how this impacts service and loyalty. Mistakes, in this programme, are viewed as positive learning opportunities.

Creating Staff Value to Create Value for Customers

Written in 1995, *Leadership and the Customer Revolution* by Gary Heil , Tom Parker, and Rick Tate is a virtual handbook for companies to help them create customer loyalty and value through the skills of their employees. The first step in skills development is constantly challenging existing action and thought. Whether a company is successful or not in keeping customers, processes and assumptions-like customer needs and expectations—must be frequently reviewed.

If we're vigilant in our efforts to question present practices and beliefs, if we bring in diverse opinions to push us outside our comfort zone and continually test the efficacy of our thinking, we may find it easier to devise a strategy without the stops and starts that have characterized so many improvement efforts in the past. Commitment requires understanding. Understanding requires learning, and learning requires persistent questioning.

Levels of performance are a direct result of present practices. This is particularly true in the skills needed to create value for customers. For example, are customer complaints actively sought and evaluated and do staff have the competency to generate complaints from otherwise silent customers? Or, are complaints considered an intrusion, something to be minimized? Are performance cycle times and causes of variation well understood, particularly from the customer's perspective? Heil, Parker, and Tate have developed a series of skill-related questions to help companies address customer-driven capability levels. Among them are:

- Are recovery efforts fast and distinctive?
- Do they proactively search for potentially dissatisfied customers?
- Are customer efforts strategically planned?

- Is process effectiveness systematically evaluated and imposed?

More specifically, are employees trained and empowered to deviate from established procedures as needed? Does staff, from the file clerk to the chairman, have the ability to identify customer needs and problems—and are they rewarded for it? Do they have the ability to use information from customers to improve processes? Do they know how to create loyal customers?

At Saint Barnabas Medical Centre, in Livingston, New Jersey, for example, the nursing staff hands out comment cards and, more importantly, calls every discharged patient to check on their experience. They also outreach to patients' families and conduct regular staff training to improve customer focus.

Several years ago, service quality educators and consultants Leonard Berry, Valarie Zeltham!, and A. Parasuraman conducted indepth research among senior executives in service corporations. They identified four gaps between the executives' perception of service delivery and the methods and tasks relating to that delivery to customers. One of these gaps was the executives' belief that their employees were unwilling or unable to meet customers' expectations.

Even when guidelines exist for performing services well and treating customers correctly, high-quality service performance is not a certainly. A service-performance gap is still likely due to a number of constraints (e.g., poorly qualified employees, inadequate internal systems to support contact personnel, insufficient capacity to serve). To be effective, service standards must not only reflect customers' expectations but also be backed up by adequate and appropriate resources (people, systems, technology).

Berry, Zeithaml, and Parasuraman identified seven factors that contribute to this gap, as well as methods to overcome them and create increased value for customers.

1. *Role ambiguity*— Employees are in conflict about providing service to customers. They need ˙clarity·

from management about what is expected, training and skills to meet and exceed these expectations, customer requirements, and an understanding of how their performance will be assessed, recognized, and rewarded.

Employees should he given updated technical training about the products and/or services offered by their company. Merck & Co., SmithKline and other pharmaceutical companies provide their detail staff and subcontractors with extensive classroom and on-thejob product and medical application instruction.

Employees should be given customer sensitivity training to develop interpersonal skills for interacting with customers and understanding their needs and problems. This is particularly true in "high touch" businesses, such as finance, hospitality, travel, or healthcare where time and money are at issue.

Airline food supplier Dobbs International Services conducted a performance perception study designed to show differences between Dobbs' front-line service staff and their customers, the airline attendants. They found that flight attendants rated performance lower than staff. This resulted in an extensive customer relations training programme for staff, a change in their title (from Drivers and Helpers to Customer Service Reps and Assistants, and new uniforms, along with business cards). Also, they set up a mechanism to monitor and reward staff. As a result staff role perception was greatly improved.

Finally, management communication skills must be such that they can communicate their performance expectations as frequently as necessary. They must also obtain employee feedback about their level of understanding.

2. *Role conflict*— This occurs when employees, either due to lack of training or lack of direction from

management, are unsure about their role with customers. For example, if an automobile manufacturer's customer support staff has been instructed to work only within black-and-white warranty definitions—even though this lack of flexibility may create owner disloyalty—this causes conflict. Further, if the same customer support staff has been instructed to cross-sell services or products to owners, even when it "feels" inappropriate, this also causes conflict. Management contributes to this by providing inadequate direction to staff; and they, themselves, may be undertrained with regard to understanding role conflict.

Conflict also occurs when customers (internal or external) queue up too fast for employees to serve comfortably. Sometimes, first-in-first-out and paper shuffling takes precedence over the real customer priorities; but staff have not been given skills necessary to make these choices.

Role conflict can be effectively eliminated through staff training: Defining service roles in terms of customer expectations, priority setting and time management, compensation based on performance quality delivery and customer loyalty.

3. *Employee-job fit*— When the match between employee skills and the level of customer delivery is poor, the company has to look seriously at its hiring and selection processes. Although controversial, this may even include assessing personality traits and characteristics.

4. *Technology-job fit*— If employees are not given the tools and training—such as updated computer language instruction—the customer loyalty impact can be as negative as poor employee-job fit.

In addition to job-specific training and cross-training, truly commitment-based companies provide self-development and extended education opportunities as well as incentive compensaticn such as profit sharing. Some companies, like Levi Strauss, are offering groundbreaking long-range bonus/salary plus compensation levels to all employees who remain loyal to the company. "Competing effectively for first-rate service providers is essential to success in a service business. Companies that excel in service select and develop employees carefully, choose appropriate technology, and concentrate on the fit among employees, technology and jobs."

5. *Inappropriate supervisory control systems* — Related to fit are the company's evaluation and reward system. Skill sets are often developed and refined more to be self-fulfilling prophecies of management's evaluation systems, rather than customer's requirements.

To be customer-driven, reward and recognition systems should be fully aligned with performance and customer loyalty. This circles back to the company's strategic motivation. When companies use behavioural and output standards for performance measurement, they must be consistent with customers' expectations. Companies such as British Airways, Publix Markets, and Federal Express provide such incentives and rewards — bonuses, profit sharing, and recognition — for outstanding service. Fort Sanders Health System, a managed care company, introduced a corporate-wide "gainsharing" programme to involve all staff in customer focus.

"A vital ingredient for excellent service quality delivery is recognition of employees' performance. Employees' performance must be continually monitored, compared with service standards, and rewarded when outstanding. A performance measurement system sensitive to high performance and tied to appropriate rewards can be very motivating, especially when workers know that others will learn how well they are performing." This also extends to teambased rewards, when staff, as a group, create or add skills that bring value to customers and earn their loyalty.

6. *Lack of perceived control*— This is the limited amount of flexibility and authority employees feel they have in making decisions that involve customer needs or customer problems.

If approval from other departments is required for a contact person to act, customer responsiveness, customer loyalty, and employee morale are all negatively impacted. Front line empowerment training, as well as decision-making authority, is important, because it serves both the company and the employee.

The more proactive and responsive companies are with customers, and the more they eliminate standardized or rigid approaches for relating to them, the faster and better employees can develop their contact and problem-solving skills.

Bowen-Scarff Ford's one-page employee manual encourages staff to do things their own way and use their best judgment. Eastman Chemical continually measures the level of staff empowerment and motivation. Jostens developed a process called ESM (Employee Satisfaction/Motivation) which links individual and group instruction and their commitment to customers. Part of L.L. Bean's lore is the customer service person who hauled a replacement canoe over several states from their Maine headquarters so the customer would have it on time for a fishing trip.

7. *Lack of teamwork*— Do company management and staff, and cross-functional employee groups, perform as a team to create loyal customers?

If employees have been trained to regard other employees as customers—recognizing that all functional areas have contributory roles in creating customer loyalty – and if employees are committed to the company as well as to customers, then there is basis for teamwork. Shell Oil Company, for one, has had extensive training for employees to improve teamwork in their relations with customers. USAA has developed an employee involvement programme called PRIDE, which creates an atmosphere of customer focus, employee empowerment, and team building. Corning has cross-functional account teams, as does Nabisco. Often, such teams mirror the account's sales structure. Teams provide support in logistics, accounting, and planning, in addition to sales and service.

"In organizations where teamwork exists, employees accomplish their goals by allowing group members to participate in decisions and to share in the group's success. Teamwork is the heart of service – quality initiatives – employees need to work together to have service come together for customers."

An example of a company that has addressed and overcome all seven of these service problems is the Ritz-Carlton Hotel Corporation. As previously discussed, front-line Ritz-Carlton staff is empowered to rectify guest problems with up to $2,000 per incident; however, they are trained to be proactive in customer relationships and service.Another example is Federal Express. With over 90,000 employees moving 1.5 million packages per day through 170 countries, training and motivation are among its highest priorities. They have a Survey Feedback Action (SFA) mechanism, begun over a decade ago. This research allows employees to express feelings about their training, management, pay, and benefits anonymously. Everyone receives extensive training:

- Call centre agents receive six weeks of technical and interpersonal skills training.
- In addition to in-depth instruction, every six months couriers, service agents and other customer contact staff must participate in a job knowledge test.

- Each person receives a personalized evaluation targeting areas which require review or upgrades.

Federal Express has provided staff with decision-making authority and an array of pay-for-performance incentive and recognition programmes. The goal is to have a quality and customer loyalty focus in every area of performance. Federal Express has also established Service Assurance and Service Action teams to identify problem root causes (late shipments, damage, lost packages, etc.) at a local level.

Customer loyalty expert Jill Griffin has pinpointed three training and reinforcement basics for having a staff that is focused on customer retention:

- *Empowerment training* — Companies like Marriott are stressing staff awareness of lifetime, or long-term, guest value. This encourages proactive, loyalty building service behaviour.

- *Product/service knowledge* — One of the reasons for Infiniti customer retention success is that *all* dealer staff, including clerks and receptionists, attend their six-day product and customer service programme.

- *Staff retention* — More than compensation plans and motivational seminars, staff need the skills, rewards, and empowerment to build the business with their own ideas. "Your employees are just like your customers. Treat them with respect and allow them to make their own decisions and they will treat your customers in the same manner. But equally important, don't tolerate in employees a casual regard for loyalty."

Motorola could easily have been an example, for they have created a culture, or style, totally dedicated to the concept of quality. It is, however, the collective skill sets of employees that sustain the culture. Motorola could also have been a paradigm example for structure, because of their emphasis on teams, or systems, because they have a Customer/Market Driven Continuous Improvement model which is built on a sophisticated information system. But it is the skill sets that sustain the system.

Part of Motorola's corporate mission is that, in selected segments of the electronic industry, they will be successful by providing their worldwide customers with "what they want, when they want it, and with Six Sigma product quality." This translates to tolerating no more than 3.4 defects per million parts. Achieving that objective, they found, necessitates a highly skilled, well-motivated, well-rewarded work force.

A member of the Fortune 50, with over 100,000 employees, and winner of the Baldrige Award in 1988, Motorola produces, sells, and services a broad array of communication, component, computing, and control products. One fundamental objective is total customer support, a stated initiative which requires participative management within, and cooperation between, all elements of the organization.

They have made training and education a corporate strategy for building competitive advantage. All employees receive a minimum of five days of training a year, totalling several million hours annually. Their training investment is around 4per cent of payroll (exceeded only by General Electric and U.S. Robotics among major companies), or $150 million annually. Ongoing training has also created a common language and focus on continuous improvement throughout Motorola. One unique result of this learning culture is T.C.S., a corporatewide competition for problem-solving teams. The competition is open to any problem-solving team — functional, cross-functional, labour, sales, service, or administration. Teams may have shorter or longer duration projects of up to six months. Direct customer or supplier involvement on the teams is highly encouraged.

Projects are selected by teams to focus on one of Motorola's corporate initiatives — participative management, quality, development time reduction, or profit improvement.

The skills required are management, relationship building, goal-setting, monitoring of and analysis of progress. In addition, the winning teams are expected to be able to train and educate others to replicate, or improve, their achievements. And, they are recognized and rewarded for their contributions. In the competition, points are awarded for how well projects are executed in several areas:

- Project selection criteria
- Teamwork/mutual support
- Analysis and recommendations
- Deployment, action, and results

The company created Motorola University, a corporate training organization for improving employee skills on a global basis. It has become the centrepiece for the company's rolling three-year training plan, in which most of the employees have individualized training programmes. Motorola University offers more than 600 courses in 14 locations around the world, and delivers more than 100,000 days of training each year to employees, suppliers, and customers. Some training is basic — math skills, remedial English, machinery maintenance — but much of it focuses on problem-solving, proactivity, critical thinking, group process, and relationship-building.

An outgrowth and extension of formal training programmes is what Motorola calls "embedded learning," on-the-job apprenticeship and mentoring. This is providing staff members with practical, experiencebased training.

Productivity per employee has risen by 139per cent as a result of problem-solving training. Motorola estimates that it has saved over $4 billion in productivity improvements through problem solving, and that each dollar invested in training returns $30 in productivity gains within three years. They are planning to provide each employee with 80 to 100 hours of training per year by the end of the decade — an annual investment of over $300 million.

Paradigm Example: DeMar Plumbing

DeMar Plumbing, Heating & Air Conditioning, a multi-million dollar HVAC service business headquartered in Clovis, California, could, like Motorola, have been an example in any one of the seven S's. They work to quoted prices, provide 24-hour/7-day service at no extra charge (even on evenings, weekends, and holidays), provide long-term financing, one year guarantees (compared to 30 days for most of their competitors), and guaranteed same day service whenever requested. DeMar calls this their Service-Based Management programme.

DeMar is corporately committed to providing the best, most valueadded service. They want long-term customers and are willing to invest to meet that goal. They do this, first, by hiring staff who will continue their standards through personal sacrifice and who fit their desired proactive personality profile. Field plumbing staff are called Service Advisors, and they are trained, supervised, and rewarded in a manner that reinforces DeMar's objectives.

DeMar spends 2per cent of the previous year's sales on Service Advisor training — technical, customer relations, sales, and interpersonal relations. They maintain a $15,000 video/audio tape library for instruction. They have created Service Advisor teams for skills building, and each team has a leader who handles most of the training, with peer reinforcement.

Compensation is commission-based, rather than salary, so Service Advisors have unlimited income potential and incentive to do well (some make over $60,000 a year). DeMar reinforces this with advertising, brochures (to customers and their neighbors), gift certificates, thank-you cards, and Preferred Customer Discount Club cards. Their pay scale is based on "happy customers," with positive points given when customers report that the work met or exceeded expectations, negative points when it didn't, and points given for orderliness of service trucks. Among reward and recognition elements of DeMar's compensation programme are:

Monetary Rewards

- $25 local gift certificate per week for best idea in service improvement, voted on by teammates (winner qualifies for end of year drawing for trip to Caribbean)

- $25 local gift certificate per week for Service Advisor with most point

Non-monetary Rewards

- Annual awards banquet for all employees

- *DeMar Star of the Year*—plaque and pin to person making the greatest contribution to DeMar (selected by company president)

- *DeMar Super Star*—plaque and personalized parking space awarded to one Service Advisor and one office person for greatest contribution (selected by team vote)

- *Iron Man*—trophy awarded to Service Advisor bringing in year's highest gross sales

- *Best "Shared Value"*—Plaque and Disneyland trip for staff members identifying how DeMar is differentiated from/better than competition

- *Top Points*—Plaque and pin awarded to Service Advisor with the year's highest total points

DeMar's schedule of monetary and non-monetary awards are designed to build teamwork and a cohesive culture, and an atmosphere where the entire company works on ways to improve service skills. They believe that positive customer perception is all-important, and all processes are focused on providing customers the best service and most positive treatment. At the beginning of each year, employees set performance goals and identify how they will reach them. DeMar helps everyone meet or exceed goals. The company has grown at an almost 50per cent annual revenue rate since the mid-1980s, even with the highest prices in its service area.

Paradigm Example: ProActive Software, Mountain View, CA

Lori Laub, formerly ProActive Software's vice president responsible for customer support, quality, training, and consulting, presented their six "people" components for developing a superior service centre at a 1993 *Inc. Magazine* customer service conference. The components were: hiring, training, compensation programmes, empowerment, rewards/ recognition, and career paths.When hiring service staff, ProActive looks for individuals who are customer-oriented and are, or have potential to be, technically capable:

- Basic technical and communications skills
- Growth potential
- Team oriented attitude and work style
- Desire to be a part of the service team in your company
- Customer orientation
- Varied educational and technical backgrounds

ProActive believes in active training and skills testing:

- Provide formal instruction
- Incorporate adult learning techniques
- Have dedicated personnel
- Create testing programmes to measure knowledge
- Have high but attainable standards
- Teach problem-solving techniques — not answers
- Provide guidelines to do what's right for the customer
- Teach customer service skills

Compensation should be motivating and value-based:

- Tiered starting salaries
- Product knowledge testing programme linked to increased salaries

- Annual compensation adjustments based on quarterly performance reviews
- Productivity/satisfaction commissions
- Sales commissions
- Tied to customer perceived performance
- Profit sharing

ProActive believes that having a value-based compensation programme provides three benefits. First, they have the flexibility to reward individual skill development as contributions are made to stated organizational objectives. Second, as ProActive's organizational needs evolve, these changing requirements can be translated into factors that stimulate individual motivation to support the changes. Finally, the compensation programme encourages staff to understand and support ProActive's strategies and plans.They also have created a professional, service-oriented culture which empowers staff to exceed customers' expectations:

- Provide guidelines to do what's right for the customer without "giving the store away" — not inflexible rules
- Hold personnel accountable to clearly communicated expectations
- Involve employees in new programme, planning, and implementation activities
- Instill a "can-do" spirit

ProActive uses staff rewards and recognition to recognize outstanding performance while, at the same time, to identify desirable skills and preferred behaviour.

- Semi-annual department banquet and awards
- Quarterly presidential awards

  Annual company-wide excellence awards

  Celebrations for industry recognitions/milestones met

An area that truly sets ProActive apart from their peers, indeed from companies in any industry, is career path opportunities for their customer service staff. In most organizations, customer service staff (like operations, accounting, marketing, etc.) receive little exposure to other areas of the company, nor does the company perceive them as having skill growth, or contribution, potential in a different function.

Their somewhat contrarian view, consistent with ProActive's philosophy of company-wide skills-building, is that the organization will be strengthened through active creation and support of career path opportunities.

- Provide challenge and growth through skill building and exposure to new ideas, processes. Provide opportunities to try new jobs before permanent transfers Require support from corporate management Require support management staff to create opportunities for their team members.

CORE VISION, VALUES, AND MISSION

Vision, as defined by management educator James Collins, is "simply a combination of three basic elements:

1. An organization's fundamental reason for existence beyond just making money;

2. Its timeless unchanging core values; and

3. *Huge and audacious* — but ultimately achievable — aspirations for its own future. Of these, the most important to great, enduring organizations, are its core values."

Vision is essentially about what the future represents for the company, its impetus or mechanism for direction. Often this is a concept as defined by the leader, or founder, who also takes on the responsibility of moving or converting the vision from aspiration to eventual reality. However, to be successful, the vision should be shared, or at least accepted, by others in the company.

In fact, forward thinking companies encourage vision and visionary conceptualizing throughout the organization. Another example is 3M, where vision and innovation are encouraged — so much so that staff members can take up to 15per cent of their working time to develop new concepts. Finally, *time* is an important element of vision. It should be far enough into the future so that it is aspirational, yet close enough that it can be used for guidance. As businesses become increasingly hightech and service based, both of these attributes need to be kept in balance.

Values and beliefs are the business of the company's direction and commitments. What does the company stand for? What is its identity? What business is it in? What are its long-term goals? Thomas Watson, the founder of IBM, defined nine core values for the company, of which three — respect for the individual, service to the customer, and the pursuit of excellence — are particularly important.

Although the core values of founder J. C. Hall (of Hallmark Cards) were considered strong — participatory management, teamwork, continuous improvement, listening to the customer, creativity, ethical behaviour, respect for the individual, and commitment to the community-after over 80 years in business it was time to restate and update them. Charles Hucker is responsible for public affairs and communication at Hallmark. He was part of a nine-person team that, several years ago, was tasked to create organizational "reinvigoration," of which values was the foundation. They crafted a statement of values, vision, and mission, called "This is Hallmark," that was clear, concise, and positive. As Hucker advises:

If you're about to begin a process of writing down your company beliefs and values, you must start with what your beliefs and values are right now. Start with what's real. Then state your goals. Use your own words to describe your organization. Don't use sweeping statements that cou..d fit any company. Your statement should be brief and crisp. And while communicating about values and ethics is essential, nothing can substitute for action.

In a recent *Fortune* magazine article, Thomas A. Stewart identified values as the "something" that brings employees together to work for a common goal:

Values don't sprout in the CEO's office or the HR department; they don't bloom on organizational trellises — the armature of boundaries and lines where the company ends and the world begins. They grow out of core professional skills, communities of practice. Here, as everywhere in business, formal organization matters less. We're flat, fluid, networked, virtual; we're an adhocracy, we work in projects and cross-functional teams. If this is where we live, this is where we will find our values. They grow where all ladders start: in the work, not the organization chart.

In *Managing the New Organization*, David Limerick and Bert Cunnington identified three types of organizational values, which they termed transcendental, strategic, and operational. Transcendental values are the core, those which built the original culture. Like the values established by Thomas Watson, they are enduring but flexible, applicable to existing (and any new) businesses, and to all organizational levels and functional areas. Strategic values can sometimes be confused with strategy itself, since these values might relate to high quality levels, such as Motorola practices, or high levels of customer service and staff empowerment, as exhibited by Ritz-Carlton Hotels.

Finally, operational values, such as dress codes, may have an impact on behaviour, but they are the least flexible and need to be changed, or at least addressed, with frequency.In *In Search of Excellence*, Tom Peters and Robert Waterman identified seven basic beliefs, or values, of the companies they had evaluated:

- A belief in being the "best"
- A belief in the importance of the details of execution, the nuts and bolts of doing the job well

- A belief in the importance of people as individuals
- A belief in superior quality and service
- A belief that most members of the organization should be innovators, and its corollary, the willingness to support failure
- A belief in the importance of informality to enhance communication
- Explicit belief in and recognition of the importance of economic growth and profits

Clearly, values are the principles that define an organization's style. For a company to make its purpose clear to employees, for it to be committed to a set of goals, to assure that all employees can be partners in the enterprise, values must be known, understood, repeated with regularity, and lived through the culture. As Charles Garfield stated in summarizing the role of shared values:

Values are the overarching principles to which an organization and its members dedicate themselves. They are the foundation on which the organization is built, the underlying philosophy that guides such things as how employees are treated; how "outsiders" such as suppliers, customers and distributors are viewed; whether and how environmental and other societal issues are addressed; how much of the budget is devoted to education and training; how profits are shared; how growth will take place; how quality is defined; and what benefits employees receive. Values shape the corporate culture and determine the environment in which employees will operate and how the organization will interact with outsiders.

A sampling of core values from well-known companies illustrates the breadth of Garfield's definition: Boeing (Being on the leading edge of aeronautics); General Electric (Improving the quality of life through technology and innovation); Nordstrom (Excellence in reputation, being part of something special); Sony (To elevate the Japanese culture and national status); Wal-Mart (Swim upstream, buck conventional wisdom); and Walt Disney (No cynicism allowed).

Honda's contains values with a very Japanese perspective ("ambition and youthfulness," "brighten your working atmosphere." "Learn, Think, Analyze, Evaluate and Improve"). Ritz-Carlton, reflecting its status as a Malcolm Baldrige National Quality Award winner, is focused on quality, empowerment and staff participation, and performance measurement. A key difference in the statements of Ritz-Carlton — and one seen in relatively few other companies — is 100per cent customer retention.

The mission statement is the declaration of the goals, or objectives, that direct the company's people and processes. Often confused with vision, mission is much more tangible in terms of determining the company's central reason for being in business, strategically and in the near-term. Militarily speaking, it's the banner behind which everyone in the company marches. It is the message the company wants to send, and share, about itself and its employees, customers, quality, personality, place in the community, and financial objectives.

Companies frequently invest many hours of senior management time and energy to craft a mission statement which, hopefully, will have true meaning for employees, suppliers, and customers. Almost as frequently, however, the statements are long, ponderous, and uncompelling, having more to do with operating efficiency, and the products' or services' financial objectives.

As discussed by Karl Albrecht in *The Northbound Train*, effective mission statements should (1) provide a definition of what customer needs will be addressed by the company, (2) identify a basic tangible value which will be provided for the customer as it aligns with needs, and (3) state the differentiating elements the company will provide to get, and keep, the customer's business. Albrecht calls these "the three components of customer-need premise, deliveredvalue premise, and means for creating value.

In 1995, *The Philadelphia Inquirer* reported on the demise of an ice cream company which had been in business for over 100 years, putting several hundred people out of work. Their mission statement had read: "The company is committed to sustained and profitable growth. We will achieve this by continuous innovations and excellence in the marketplace, coupled with a conspicuously lean and costconscious management style and minimal waste." This is a statement which, essentially, says nothing about the company. There is no customer-need premise, no delivered-value premise, no means for creating value. It doesn't even identify the company's business.Compare this to Gerber Products Company's corporate mission:The human, physical and financial resources of Gerber Products Company are dedicated toward:

- Establishing Gerber as the premier brand on food, clothing and care items for children from birth through age three.
- Giving our customers and consumers what they want all the lime, every time, on time.
- Continuously pursuing improvements in all phases of our business.
- Seeking intelligent risks that will build shareholder value.
- Providing long-term shareholders with superior returns.
- Creating opportunities for all associates to achieve their full potential.
- Maintaining mile Gerber heritage as the authority in the field of infant and child nutrition and care.

This statement is extremely long. Much of it, between the first and last bullet points, set forth no value; but, at least, value and distinctiveness are expressed. Gerber might have been better served just to use the opening lines, the first bullet point and possibly the last. This would have conveyed, succinctly, their value message.

Albrecht gives his perfect example of a company mission statement: Po' Folks Restaurants, a country-style Southern food dining chain headquartered in Mount Pleasant, Kentucky. "We always want to be the friendliest place you'll ever find to bring your family for great tasting, homestyle cooking, served with care and pride in a pleasant country-home setting at reasonable prices." This mission statement has equal meaning for employees, customers, suppliers, and the financial community.

It is brief, defines the company's value message, provides an operational foundation for staff, and certainly identifies the company's unique qualities. Albrecht summarizes his five criteria for an effective mission statement:

- *Definitive* — The mission statement should leave no doubt as to the nature of the customer, the value provided by the company, and how they conduct their business.

- *Identifying* — There should be no question to the scope, or line, of products and service provided to customers.

- *Concise* — Missions, even if combined with an expression of vision and values, lose their impact if they begin to take on the length of the Gettysburg Address. They should, at most, be a very simple paragraph. At least, they should be capable of being said in one breath. Po' Folks passes this test, Gerber does not.

- *Actionable* — It should tell the reader how the company delivers value.

- *Memorable* — Like Ford Motor Company's "Quality Is Job # 1," it should be associative for all of the company's constituency groups. While Ford's mission statement does not precisely meet Albrecht's other four criteria, it does express the spirit and mission of Ford's corporate effort. Also, it's easily related to their business.

The central importance of mission, vision, shared values and superordinate goals is the sense of meaning created for employees and customers. Unlike strategy, structure, and systems, this is one of the soft S's of the Seven-S Framework — which offers focus and ideals for the leadership to represent, and a basis for the company's culture to grow and develop.

EMPHASIZING CUSTOMER LOYALTY

Loyalty programmes are a necessary part of a successful CRM programme and long-term customer retention. Compare available programmes, learn what a loyalty program can achieve for your business and gain an eight step guide to designing your own succes. In customer relationship management (CRM), customer life cycle is a term used to describe the progression of steps a customer goes through when considering, purchasing, using, and maintaining loyalty to a product or service. How many times have you heard that it costs several times more to attract a new customer than keep an existing one?

Or that satisfied, loyal customers become more and more profitable over their lifetime as they purchase new products, updates on old ones, supplies, and services, all the while recruiting others to do the same? In fact, that customer loyalty is the single most important driver of growth and profitability? When several of us first began researching these relationships, findings such as these were news. Today we take them for granted.

The economic and cultural benefits of keying on customer loyalty and value cannot, ultimately, be achieved unless shared values and overarching goals are the hub of the company's wheel of enterprise. As Peters and Waterman identified shared values, they are the significant meanings that an organization communicates and imbues in its members.

Churn rate is a measure of customer or employee attrition, and is defined as the number of customers who discontinue a service or employees who leave a company during a specified time period divided by the average total. Some companies, such as Great price, a supermarket chain, have a statement of philosophy that is commitment-based, defining its complete focus on customers. Others, such as Heavenly, a consumer products company, and German-American, a beer company, say very little about their commitment to providing value for customers. This doesn't necessarily mean that the companies don't consider customers important; but in their public statements of beliefs, it indicates a relatively low level of priority.

Two companies in the same industry — oil — have radically different customer priorities. State is committed to customers, as reflected by its Principles and Objectives: "To deliver to customers only products or proven high quality at fair prices and to serve them in such a manner as to earn their continuing respect, confidence, and loyalty, both before and after the sale."

Textile manufacturer Milliken & Company's customer goal statement is rather middle-of-the-road regarding customer value: "To provide the best quality of products, customer response, and service in the world through constant improvement and innovation with a bias for action." Where Milliken's values translate to complete customer commitment, however, is in its quality policy (signed by CEO Roger Milliken) and its statement of seven business basics:

Quality Policy

Milliken & Company is dedicated to the continuous improvement of all products and services through the total involvement of all associates. All associates are committed to the development and strengthening of partnerships with our external and internal customers and suppliers.

We will continually strive to provide innovative, better and better quality, products and services to enhance our customers' continued long-term profitable growth by understanding and exceeding their requirements and anticipating their future expectations.

Business Basics

1. Quality (Products, Services)
2. Close to the Customer (Closer than Competitors)
3. Innovation (Change for Better)
4. Value (Cost, Quality)
5. Specialization (Best, Not biggest)
6. Bias for Action
7. Partners for Profit (Customers, Suppliers)

Milliken reinforces these statements with what it terms "people values," which include quality in selection, education, appraisal, safety, and career development; integrity, both an expectation and a necessity; teamwork, effort and results; strong communication and "management by walking around"; hard work; and recognition opportunity for all individuals and teams. Thus, through its expression of commitment to customer relationships and the provision of lasting value, Milliken has also created customer loyalty strategy, structure, systems, staff, style, and skills.

In *Built to Last,* James Collins and Jerry Porras expressed their belief that values — core ideologies — have to be authentic: "You can't fake an ideology. Nor can you just intellectualize it. Core values and purpose must be passionately held on a gut level, else they are not core." Milliken is an excellent example of a company that has purposefully interwoven customer relationships and partnership into its very fabric.

Banc One is another. Part of their core ideology is something called Uncommon Partnership. Essentially, it is the focus of all resources on customer commitment and the provision of value to them. Like many other companies, they have experienced some staff reduction; but, as the seventh largest U.S. bank and, with 69 affiliates in 11 states, they have worked hard to live their mission and values.

Companies that are passionately focused on customers and infuse that passion within their employees, so that the employees share and own the same passion, will realize all the rewards that being at the top of the Customer Loyalty Pyramid^{SM} has to offer.

Companies that include "customer satisfaction" in their statement of vision, sales, and mission are, conversely, sending a message to employees and all constituencies that they are interested in their customers, but not passionately committed to them. A part of the NCR (now AT&T Global Information Systems) statement reads: "Because quality and customer satisfaction are viewed as one and the same, our near-term quality goal is to achieve world-class customer satisfaction." The statement goes on to say that "there will come a time when all operations should be defect-free, and when every employee can 'guarantee' his or her customer's satisfaction.

Thus, our supreme goal is 'guaranteed customer satisfaction'—doing the right things right, the first time, and every time!" Not only is this rather florid and stilted, but it likely has little meaning to either employees or customers. Like many satisfaction-based and performance-based companies' statements, it emphasizes staff delivery of customer satisfaction. Customers want value.

If companies are ideologically and actually committed to customers, like Milliken, then this should be reflected in believable statements of vision, value, and mission. The following excerpts from published company statements provide windows into their perceived stage, or level, of customer commitment:

Satisfaction-Based/Performance-Based

"The Corporation is committed to providing innovative engineering solutions to specialized problems where technology and close attention to customer service can differentiate it from commodity production or job-shop operation. To deal fairly with customers, employees and suppliers in order to merit their continuing patronage and support.

Every customer is entitled to the timely receipt of the product or service he buys from us at a fair price. Every customer is entitled to receive the quality level he requested or that which is consistent with what we ourselves would expect to receive. Every customer is entitled to receive a safe product that will not maim or injure him. We must excel in meeting the needs of our customers for prompt and efficient service. A Squibb customer must be a satisfied customer.

Customer First—We deliver our service in a manner that causes our cus tomers to perceive value-added. We will know their preferences and per ceptions and market our products and services accordingly. We will judge our performance against the standards of the customer. Only by satisfying our customers are we able to provide an attractive return to our stockhold ers and opportunities for our employees.

Quality is the key to our future success. MetLife customers are our first priority: without customers, there is no reason for a business to exist. Exceeding the expectations of our customers will make MetLife 'The Quality Company'.

Commitment-Based

We are also committed to providing the best customer support in the indus try. This includes: meeting customer needs quickly, interacting with customers professionally, focusing on uptime, and offering a complete range of services. In this manner, we strive to earn the loyalty of our customers. We believe our first responsibility is to the doctors, nurses and patients, to mothers and all others who use our products and services. Outstanding customer service has become an imperative for superior performance, especially when poor service can damage brand equity.

A recent study finds marketers are aware that data breaches can harm their brand, but few have a plan of action in place. Hotels need to be direct in dealing with the problem.This provides strategies and tactics that companies can use to consistently deliver their customer service promise. While marketers are making customer development a priority, they have a significant disconnect with the realities that drive effective customer targeting, acquisition and retention.

Committing ourselves to integrity, it is our purpose to: I. Earn the respect, confidence and loyalty of OUR CUSTOMERS by serving them so well that they profit from their association with us. Our company exists primarily to serve the customer.

Without the customer and his need for our products and services we have nothing. We believe it is absolutely necessary to anticipate our customers' needs for products and services of the highest quality. Once a commitment is made to a customer, every effort must be made to fulfill that obligation.

"The company that fails its customers, fails! We will be superior to our competitors in providing the highest value to our customers at a fair price. We will constantly listen to our customers, respond quickly to their current needs and anticipate future needs."

"Customers are the focus of everything we do." (Owens/ Corning Fiberglas, Guiding Principles)

Some companies are moving up the pyramid, from satisfactionbased or performance-based to commitment-based. They straddle a customer satisfaction/customer loyalty and value fulcrum, or will straddle it until they have reached the conclusion that customer satisfaction is a past, less compelling or effective operating paradigm. An example is Electro Scientific Industries. A portion of their Statement of Purpose reads:

Customer satisfaction is what keeps us in business and makes growth possible. "Value" is the key — a concept that includes not only product usefulness and quality, but also price and performance and partnership. We believe in providing unmatched value. That's accomplished by meeting our customers' needs better than anyone else.

We must be customer oriented in all we do. Every job at ESI exists to satisfy our customer's needs. We do this by developing, producing or supporting products and services for our customers.

Our products will be judged in terms of our ability to meet our customer's need for innovation, quality and reliability. By meeting those needs we are giving our customers real value.

Another example is the Oldsmobile Division of General Motors, whose mission statement reads: "The Oldsmobile Team will work together to earn and keep customers by providing internationally focused vehicles and uncompromised satisfaction throughout the shopping, buying and ownership experience." A third is the mission statement of Dean Care, a Madison, Wisconsin HMO's customer service department: "We are working to achieve excellence in customer service by going the 'extra mile' to provide quality and educational information to our external and internal customers.

In all three cases, customers and staff are given rather mixed messages. The companies want to keep customers and provide value; however, they are endeavoring to do so through satisfaction or quality. So, while excellence and customer loyalty are the stated goals ("earn and keep customers"), satisfaction ("meeting customers' needs") is the intended mechanism. As has been discussed, reactively satisfying customers will not differentiate one supplier from its competitors, but putting in place a proactive commitment-based culture and infrastructure, or sustaining value leadership, will.

Over the past few years, there has been some disagreement among management consultants, management scientists, and management theorists about the real worth and meaning of mission, vision, and value statements. Looking outward at customers and inward at employees, they provide companies with the opportunity of defining themselves and from that definition to create a culture (style), capabilities (staff, skills), direction and infrastructure (structure, systems, and strategy).

Perhaps the most profound statement on the importance of shared values has come from noted behavioural psycholgist Abraham Maslow. In a little-known 1965 book on organization development, entitled *Eupsychian Management*, he reasoned that if everyone in an enterprise is clear about the enterprise's vision and goals as well as its values, practically everything else — operations, internal and external relationships, and so forth — can be easily addressed.

If, on the other hand, vision, goals, and values are confused, conflicting, ambivalent, or only partially understood, very little within the organization is likely to function well. As Thomas Stewart concluded in assessing values that work: "Human beings want to pledge allegiance to something. The desire to belong is a foundation value, underlying all others." In the context of customer and staff loyalty, allegiance and partnership are synonymous.

Chapter 6

Work Load Modeling

SETTING PRIORITIES AND LOCATING BOTTLENECKS

"A little bit of graphic preplanning yields large dividends."

Competition among service providers must eventually be won by the supplier who offers best price, highest quality, and most timely service delivery. These are all operations issues. Managing operations effectively is the difference between success and failure in service industries. Those operating dimensions that determine service quality will be extensively analyzed in later ones. In this, we will look at some techniques that are useful for improving work flow, reducing delay and generally getting the most utility out of available equipment and labour capacity. These techniques fit under the headings of work load modeling, priority setting and location of bottlenecks.

WORK LOAD MODELS AND PRIORITY SETTING

Modeling work load in a service operation picks up where forecasting leaves off. In some ways, work load models can be used to supplement forecasts. A good forecast of demand by service skill category, for instance, should include best-, worst- and most-likely-case scenarios. A model of available service capacity can then be employed to test for capacity overload and slack. Points of overload are potential bottlenecks where service quality may be stressed to the breaking point. Heavy slack may point to the need for more skill cross-training or more flexible working hours. Modeling the impact of potential high and low demand on the service organization's work centres permits intelligent planning for business contingencies.

Exhibit illustrates how forecasting for various mixes of business in a small law firm might be modeled against available capacity. Five major categories of legal specialization are used here as the basis of the forecast and as rough indicators of differential legal skill. The forecast offered covers one fiscal quarter (three months or thirteen weeks) of calendar time and assumes a starting level of thirty-two legal professionals working for the firm. Variation in level of business activity from best to worst case is in the range of + 12 per cent to - 15 per cent. Assuming a normal forty-hour work week and 100% billout of hours, approximately 500 hours of billable working time (using round numbers) are available from each lawyer in a thirteenweek quarter.

Hours of billable time are forecast and cumulated for the quarter by the five categories of skill/activity, then divided by 500 to obtain an approximate count of legal talent required to cover the anticipated business. Something between thirty and thirty-nine lawyers will be required to cover the forecast, depending on variation from the most to least likely level of anticipated business. For a business where the standards are more precise, the calculations might reflect added precision. Here and in many similar cases, round number approximations will serve the need adequately. Anything more suggests a degree of precision in forecasting and planning that is illusory.

Testing a forecast requires construction and use of a viable model to estimate how future circumstances are likely to stress available capacity. Modeling permits us to go beyond mere gross numbers of people required to cover the expected business demand. We may now identify the qualifications of current staff by skill/activity area—one person may be qualified in one, two or more skill areas—to determine what the effect of a best-case quarter will have on available skill capacity. The first question asked is: If the firm is lucky enough to get all the business that is possible in this period of time, can it handle the load?

Assuming that all personnel are qualified to perform basic legal research, any of the total firm strength of thirty-two can be assigned to this category. This is the maximal flexibility skill area. Entry-level law clerks who are yet unspecialized can be assigned here to cover the base requirement (i.e., worst-case demand, for instance), and anyone otherwise unapplied can pick up the remainder. The most highly skilled (and capacity constrained) categories are trial work and corporate law. In these areas, we find that the best-case business forecast slightly more than covers trial law, while corporate law is shy one full person.

Overtime might be used to cover the excess for now, but as business expands further, it may be necessary to hire or develop qualified talent in both these areas. Real estate and criminal practice are both adequately covered by available talent as long as other areas of need do not suck up the excess in available talent in either of these areas.

With an assumed current thirty-two person legal staff, best-case forecast conditions will increase the work load by about 20% if no new additions are made to the firm. This represents one extra full day of work per firm member per week. Worst-case, on the other hand, will result in about a 6% underutilization of available talent. For both most-likely and worst-case forecasts, available talent will cover expected demand for service suitably.

The modeling process goes beyond mere revenue goals for the business. It compares capacity with anticipated demand to identify gaps or excesses that can either impair service quality or raise cost. These are the critical operations management issues for every service industry. Modeling in some form or other is fundamental to their analysis and resolution.

Beyond forecasting, a comprehensive model of the business' capacity for operations is a useful way to test the impact of incoming business demand on the available skill and work centre resources. As each new customer demand is placed on the system, it is entered into the model to determine what effect the added service demand has on available capacity. Exhibit shows, for instance, how available capacity in a small appliance repair business might be affected by the mix of incoming repair orders. Eight different categories of appliance repair are covered by this business.

Maximum daily capacity for service can be determined by making reference. It is assumed here that the average repair time is one and one-half hours, with an average travel time to the job of one-half hour. Each job entering the order queue has an average potential of two hours of committed work and travel time. Most appliances are specialized to a single repair technician. Refrigerators, electric ranges and electric ovens, though, are each covered by the talents of two technicians. Each technician has at least two appliance repair skills. This permits some limited flexibility in the assignment of incoming jobs to specialists based on schedule availability.

On the day illustrated in this example, the work load is near maximum: 87.5% (thirty-two normal available hours versus twenty-eight scheduled hours). Exhibit 9-3 shows that two technicians are fully loaded and scheduled to work a normal eight-hour day. Any further work orders in these two areas must either be covered with overtime or deferred to the day following. A surge in orders on the day following could create more overtime or diminish customer service seriously. The question becomes one of whether to incur the added cost of overtime to serve additional customers today or risk impairing service standards by deferring jobs and overloading the schedule tomorrow. Again, the operations tradeoff is between cost and quality of service. Modeling makes the options clear and assists in making the best possible decision.

As the foregoing illustrations suggest, there is no requirement that modeling be difficult or elaborate. In some businesses, it might be handled through a physical model. An advertising firm, for instance, might use a pictorial layout of the firm with each work area sized to represent proportional capacity. As work comes into the firm, the surface of each area involved in a job is physically covered to reflect the work capacity committed to that job. As each job is finished, the cover representing that job is removed. A visual, real-time picture of work loading is thereby offered, which points to delays and bottlenecks that may then be managed by subcontracting or reallocating available labour skill.

Indeed, the appointment schedules that characterize operations management control in offices of every kind, including those of physicians or lawyers, and personal services like beauticians or counselors, are, in reality, models of available capacity that can be used to allocate capacity as fairly and fully as possible across current customer demand. The salesman's appointment book is a capacity (time) allocation modeling tool. Models on a simple level are already widely and effectively used to manage service operations in many industries.

More complex capacity analyses may require a computer model, which need not be overly complex in itself. Indeed, the simplest and best model is likely to be one that is implemented on a personal computer using standard electronic spreadsheet software such as Lotus, Symphony, Quatro or Excel. Even capacity that must be simultaneously modeled for both workstations and labour time can be effectively and efficiently handled with no more than these basic tools.

Flexible Capacity and Bottlenecks

In the examples just discussed, most capacity constraints are a function of the need to utilize skilled labour efficiently. Indeed, that is where most bottlenecks will necessarily occur in a service economy. Rarely, equipment will also be a major capacity constraint. In a small medical facility, for instance, it might not be cost-effective to duplicate high-cost X-ray or NMR equipment.

Larger facilities might take advantage of their economies of scale to duplicate some kinds of expensive diagnostic equipment, but even here there will be basic cost constraints and efficient scheduling of bottleneck capacity may be more cost-effective. Sound operations management methods will then be the difference between high- and low-cost service.

It is of equal importance, though, to build maximum flexibility into the system in the form of versatile equipment and multiskilled service providers. Too much specialization of either is a potential source of increased cost. Specialization is an old, deeply scored habit of operations management in the commodity operations arena. It is a habit that is easily carried over to the service sector, where it can only add cost and create inefficiencies.

Effective management of service operations requires that specialties be tolerated only where unavoidable, and that even unavoidable specialties be challenged as threats to cost. In many circumstances, for instance, it is unthinkable for a physician to perform technicians' work. That is a "waste" of skilled time and talent, even when the physician's time is being wasted anyhow. If efficiency and service are the objective, though, it must become "thinkable."

The restaurant owner who refuses to bus tables when customers are lined up out the door can only reduce his own profit by wasting opportunity for effective application of his time. Moving to cover bottlenecked work is natural and inevitable in a service environment. But expecting workers to shift downward in the status scale to accomplish needed work is likely to bring disappointment.

Moving down on the status ladder would be seen as a diminishment of one's value and status rather than as a contribution to organization effectiveness. It is often easier to cross-train technicians in skills that are on a comparable level of status. If moving down is hard, moving up without sanction is likely to be dangerous.

The technician who practices medicine or the busboy who takes orders from waiting customers is likely to be censured by the legitimate status holders. But it is poor operations management to tolerate such attitudes. It is a source of waste to fail to cross-train and upgrade talent wherever it is possible to do so. It is arrogant to refuse to do low-status work that has become a bottleneck to customer service. Flexibility in all directions is fundamental to service operation efficiency. Every opportunity to enhance flexibility must be seized.

Upgrading into higher priced, unutilized capacity is standard in the hotel and airline industries. Late arrivals at the hotel reservation desk who cannot be accommodated in the class of room reserved can easily be upgraded into higher priced quarters that are empty without increase in reservation price. Tourist-class airline passengers who are too late to get seats can be upgraded to first class. Reservations of any kind that are oversold can be compensated with free service at another time if there is no flexibility to current capacity. These are common devices for dealing with capacity inflexibility. Luxury service or accommodations are priced to break even well below full capacity, so excess capacity in luxury domains is employed to cover excess demand for the "popularly" priced offering.

Opportunities for flexibility exist in many places when they are actively sought out by the service operations manager. They must usually be identified in advance of the need, though, when the pressure of time to meet customer need does not impede search for them. That is often the difficult part. When capacity is stressed to its limits it is obvious that help is needed from some source, but the time is not available for a comprehensive search to locate the right source. The quick fix is more likely.

If it is good, it will be remembered and reused. If not, the search for better may be postponed until the next crisis. Old habits of operations management assume that departures from standard procedure are always temporary inconveniences. The new perspective on service operations management must treat them as important assets in the arsenal of service flexibility. Continual search for new and more powerful sources of adaptive flexibility that solve temporary capacity constraints is the very soul of good service operations management. Deeply grooved routine is the enemy of service operations efficiency.

Every decision that pertains to equipment and labour capacity should be made with one eye on the fundamental need for increased flexibility in service capacity. High-cost, leading edge technology that is narrowly specialized but more efficient than cheaper, general-purpose equipment is still a high-risk opportunity that should not be approved without a very large margin for error in costing.

On the other hand, equipment that increases flexibility at equivalent cost is likely to become an asset to capacity constraint problems. When the time and opportunity are available, it is imperative that experiments in capacity flexibility be designed and carried out by service operations management with all forms of equipment.

Similar experiments with labour flexibility are essential to planning for capacity crunches. Slack time in the schedule can always be productively applied to upgrade training of workers. Multiskilled service providers who can be reallocated as demand changes offer superior customer service *and* enjoy greatly enhanced job security. Opportunities to learn, practice and demonstrate command of alternative skills must be aggressively exploited. Pay structure for alternatively skilled workers must then be restructured to recognize multiple skills.

Self-service offers potential for increased capacity. Automatic teller machines (ATMs) are employed by many banks to permit twenty-fourhour access to banking services. They also offer overflow capacity service during normal working hours when capacity is stressed. On-line networking with personal computers permits near real-time information on stock market exchanges. Computer input to sales and buy orders is technically feasible. Permitting customers to directly place, sell and buy orders on their own personal computers is a device that overcomes the broker's phone line bottleneck that occurs in heavy trading times. To the extent that such self-service devices please customers, they may occasionally represent an improvement in service, despite the absence of personal contact.

Any service business that can arrange it should always have backup personnel available on-call. Fire and police services already operate routinely on this basis. Supervisors in some industries fill the capacity gaps. Banks, hospitals, communications services and maintenance routinely expect their supervisory staff to cover excess demand. In those rare situations where higher management takes the opportunity to work side by side with line operations personnel, willingness of management at all levels to get their hands on the real work is increased. Awareness of opportunity for other flexibilities is often also enhanced.

In those organizations where top management never leaves its ivory tower to engage the real work of the organization, narrow, inflexible skill must prevail throughout. In an on-call service situation everyone must be flexible. Even in those organizations where work flow is relatively predictable, an organization that is pervaded by skill and task assignment flexibility will be the competitively stronger one. Reallocation of time and talent is a primary resource for improved service operations cost efficiency and service quality.

Service organizations like Rotary, Kiwanis and Lions all know the power of flexibility and willingness to serve in any role. None permit leaders to serve longer than a specified term, usually a year as president, for instance. All have past presidents, presidents-elect and presidentsaspirant as sources of leadership in times of need. Illness or business exigency need never impair the organization's mission effectiveness. There is always alternative talent and backup experience ready to apply. Ample mechanisms exist to assign legitimate status and authority in the organization, but no one is indispensable. Everyone is available to meet the challenges of the moment.

Traditional business organizations with their relatively rigid hierarchies and status distinctions diminish role flexibility among their members. The more inflexible the business becomes, the less adaptable and the less competitive it is. Under stress these role and status distinctions become major liabilities. In the military, status distinctions between officers and enlisted personnel serve to maintain stability in peacetime. In battle where death can become the ultimate leveler of status, they are only old habits that assure clarity and continuity of communication under fire. Skill is what counts at every level, and anyone who can't fill alternative jobs is a liability.

The lessons of service flexibility are widely available but just as widely ignored within the context of standard operations method and policy. Specialization in many service industries and roles increases rather than decreases cost. Status barriers between organization levels don't improve performance, they waste manpower and talent. Flexibility and variety, rather than being the enemy of cost-effectiveness as they are in the commodity production environment, are its foundations in service. Flexibility is indispensable.

Monitoring for Bottlenecks

Capacity constraints – alias "bottlenecks" – can occur by chance anywhere in an unscheduled system, even one where capacity utilization is, by design, low. The best solution to capacity constraints is still the maintenance of a generous level of overcapacity in the system. But sudden surges in demand either overall or in some limited corner of the system, caused by factors such as equipment breakdowns or absent personnel, can overload capacity or demand an operations management fix. Where additional capacity cannot be brought on line, the recommended fix is to expedite jobs that can fill the overloaded workstation or labour capability quickly and fully to assure that the available capacity at that point *is not wasted*.

If the bottleneck is not noticed, though, that will not occur. A high priority on every service operations manager's list of duties, then, is to maintain a continuous watch for emerging bottlenecks. It is accurate, indeed, to maintain that the core accountability of any competent service operations manager is to manage bottlenecks effectively.

Effective management of bottlenecks will usually require that all incoming tasks be preallocated across available capacity to monitor demand on a workstation-by-workstation, worker-by-worker basis. Managing service operations thus requires that the task time requirements of arriving work be competently estimated and added to a current work loading estimate. It is precisely this kind of monitoring that permits potential capacity constraints to be forecast so that work can be appropriately expedited into them.

Where possible, large jobs that are on their way to or through the system and that will probably produce a bottleneck should generate immediate priority for movement through the system. Other jobs in the system that could be put through the impending bottleneck to utilize available capacity before it becomes loaded must also be expedited so that they will not needlessly end up delayed by filled capacity at the bottleneck.

In a hospital emergency room, for instance, it is highly desirable to have notice from the police or ambulance crews of a major disaster—a multiinjury vehicle accident, or a large fire in a crowded location—so that capacity can be expanded and patients en route to that capacity can be expedited through in advance of incoming crisis patients. Any incoming patients who will tie up a larger-than-usual number of medical staff signal the need to use that staff capacity *now* for any patients in the system. It may seem counterintuitive to load up capacity that is about to be overloaded, but merely letting that capacity sit idle in wait is a grievous waste when there is work in-house that could be expedited to that station in advance of the surge.

Any forecast that suggests heavy demand for service probably signals the need to begin giving priority to work scheduled into the area of heaviest expected demand. The flight that is departing for a bottleneck air terminal should be moved out first. The train that must use a track corridor restricted by track repairs must be given priority over the one that does not need that corridor. The legal matter that requires intensive team negotiation should be given priority for research and investigation. The insurance policy that requires an unusual amount of actuarial investigation should get priority through the clerical stages of processing and any other work in-house requiring actuarial review should be expedited to that area *now*.

The working principle of sound service operations management is expediting to load the potential capacity constraint. Once capacity is overloaded, there is little to be done except work it off quickly. Any jobs that must flow through one bottleneck to another are obvious and natural priorities for attention at the earlier bottleneck. Those remaining nonconstrained capacities in the system will largely take care of themselves on an as-needed basis. It is the bottlenecks that raise cost and diminish customer service. They demand first priority in attention.

Bottlenecks are always identifiable when waiting work piles up in front of the workstation. Long queues of waiting work, like long lines of slow-moving traffic, mark the obvious bottlenecks. Once the waiting work has piled up in front of the workstation, though, it is too late to take remedial action in the form of scheduling. The best that can be done is to increase capacity at the bottleneck and/or expedite work through the bottleneck to the next bottleneck—assuming you can identify the next one. The trick in sound service operations management is to anticipate bottlenecks *before* they announce themselves with pile-ups of waiting work.

That can be accomplished *only* by modeling the work load in real time to forecast the impending increased demand. That requires some form of graphic, tabular or computerized system like those illustrated earlier. Only when overloaded capacity is identified is it possible to expedite the right jobs through the system to fill the limited capacity early and fully. Any other approach is a crap shoot. Managing service work flow in a complex flow environment means modeling flow against incoming work. That is an indispensable element of cost efficiency and customer service in service operations.

Services That Satisfy Fundamental, Common Human Need

The second and perhaps most common source of service quality from those offered is that which describes the most common dimension of service: service that satisfies our basic human wants and needs. This is the sense of service that is rooted in the Latin term *servus*, referencing slavery.

In ancient times before indoor plumbing and running water were engineered, the functions of food preparation, personal hygiene and sanitation occupied the larger part of a day's activity. Social and professional pursuits were impossible without the aid of a servant. In its original form, slavery was as much a matter of task specialization as it was of subjugation. Slaves saw to the physical basics of their masters' lives while the masters attended to politics, professions and social amenities.

At the highest levels of social and economic order, slaves served in professional roles to kings and potentates. As much as anything, slavery defined continuing work obligations of specialists who supported a stable social order. Contradictory as it may seem, it is likely that Athenian democracy could never have come to flower without the assistance of slaves.

Slavery in America was distinguished by the clear racial lines between slaves and masters, which made mobility out of slavery, an ancient characteristic of the institution, something between difficult and impossible. Coexistence of democracy alongside of slavery in the New World was a contradiction in ideals unless the slave race could be defined as something less than fully human. That was the fundamental pathology of American slavery. Its forms were otherwise faithful to the traditional structures of slavery, which sharply and cleanly specialized the task assignments of masters and slaves to support organized industry.

Modern factories have preserved specialization at a level that is sometimes described as wage slavery—subsistence pay for dreary, unskilled labour and little option outside destitution for those unwilling to accept their role. The consequence of these various institutions of specialized, basic unskilled labour within a hierarchy of private or public power is a tradition of assigning low status to those who serve our basic, menial, personal and economic needs.

Basic, indispensable health support services like those of a hospital orderly, rest room caretaker, trash collector, or restaurant dishwasher are dismissed, ignored or scorned as worthy only of low-status citizens, undeserving of our respect, instead of knowing respect and appreciation for the great value of their specialized contribution to society. Service at the base of the economic structure that frees us to pursue our own political, professional and social objectives passes unrecognized for its great value.

Good service, as a result, is often taken for granted. Its absence generates complaint, but its presence merits no special notice. Where it involves professional know-how or skill, it is common for the servant to adopt a superior approach to a client as antidote to loss of status in the servant/master relationship. Personal services in luxury hotels, restaurants, beauty or barber shops, health spas, and so on, are supplied by willing, highly paid personnel who enact the role of servant for the entertainment of the customer.

It is not always an easy role to play. In the most luxurious of accommodations where servants are sometimes better paid than masters, it becomes hard at times to know who is heaping the greater volume of scorn—customers on servants or servants on customers. The provision of personal service in these settings easily becomes a contest for status. As a result, the price of personal service is high and continues to escalate.

Cost-effective personal service is almost a contradiction in terms in the current age. It can, though, still be found. Telephone and postal services are still cost-controlled offerings to the public. Libraries and fast-food restaurants supply their offerings at low cost to the public. Police and fire protection, sometimes supplied on a volunteer basis, are a bargain in many communities. Complaints about quality in these services are often tempered by concerns over cost. Services of the Red Cross in disasters and of local medical emergency personnel are often performed at no cost to the recipients. Food and shelter for the poor or indigent are supplied at minimum or no cost through various community agencies, often supported financially by voluntary donations. A variety of fundamental human services continues to be offered on a cost-minimized basis in many corners of the economy.

A tradition of service without profit to one's fellow man continues through churches, service clubs, lodges, fraternities and sororities in every community. The Christian ideal that "the last shall be first and the first shall be last" holds sway in service projects targeted for community benefit. But the deep, gut-level tradition of low status in the role of service-provider continues to dominate. Where it does, the servant/master relationship often looks more like competition for the upper hand.

It is an enigma of current times that much personal service comes at moderate to high cost but that the quality of service given is often low in the sense of the service supplier's respect for the customer. Insurance companies often treat customer claims as if they must start with an assumption of attempted fraud. Banks, hospitals, physicians, dentists, lawyers, credit-reporting companies, public transport, personal service and restaurants too often deal with the customer as a faceless object to be controlled and exploited rather than as a person to be served.

The quality of these services is seriously undercut by the double whammy traditions of mass commodity operations methods and assumed low-status conflict of service provider.

There is nearly unlimited opportunity in these industries to create and deliver high-quality personal service if only the effort will be made.

Indeed, future competitive success in these industries will most likely depend on cost effective delivery of the highest quality of personal service. High quality at low or reasonable cost is the most consistently successful business strategy known. Applying it, though, requires exceptional management of service operations to assure consistent delivery of courteous, individualized, respectful service to every customer.

Entertainment as Service

The third and concluding dimension of service is that of entertainment. In modern Western culture, entertainment has taken on the character of superficial distraction through passive spectatorship. While that may be one of the dimensions of entertainment, mere passivity is not the true core of its meaning. Entertainment in its most fundamental form requires a bond of basic and vital communication between entertainer and audience. The etymological root of the word is in the French *entretenir* meaning to hold between or bond together in some fundamental way. Passive distraction or interest is certainly encompassed within this meaning.

But so also is the most passionate of love affairs. Those who entertain one another reach into the deepest levels of awareness to communicate. It is instructive that the championship football team, consummate actor or riveting comedian who succeed in entertaining us seem to enact our most basic aspirations for high human expression. For those who have lost either the capacity or the opportunity to articulate their most urgent needs, entertainment in the passive sense is life's most compelling involvement.

But entertainment is also discovery, learning, accessing experience we didn't know existed to meet life's opportunities and crises. We may entertain ourselves or we may be entertained. Entertainment from extrinsic sources is an aid to self-discovery and growth. Drawn from intrinsic sources, it is the creation of a higher self. Entertainment as service is on a par with sorcery. It has the power of magic and the uplift of spiritual enlightenment. The linebacker on our favourite football team who breaks up the opposing team's potentially winning pass enacts our own struggle to win against stiff, competitive odds. The actor who plumbs the depths of personal agony in simulated tragedy echoes our own pain. The comedian who amusingly utters the acid truth about the powerful expresses our own unspoken dismay. Were we willing to discipline our own actions to the challenge, we might forsake passivity in favour of our own active expression. But it would require a wholly different approach to expression. We would have to actively entertain life at its source with all the risks and rewards that entails.

Television, movies, theater, even amusement parks and a walk on the city streets can supply passive entertainment. But there is also powerful entertainment in the services of a skilled physician, an experienced lawyer, an inspired teacher or a good local library. Indeed, the highest source of quality in these services may be their power to entertain us. To the extent that the customer insists on discovery over distraction, they may become entertainments that transform.

The quality of an entertainment is, unfortunately, not found in the customer's activity or passivity toward it. Passive entertainment is as powerful in its impact as an active form might be and sometimes more so because it makes lesser demands on us. In some ways, indeed, it is more satisfying in that it requires no great personal discipline or sacrifice to participate passively. The investment to become an active performer could be immense. The physical conditioning and hazard of injury demanded by the play of football, the talent and training required of a successful actor, or the timing and wit needed in the delivery of a comic's laconic lines-none of these need be mastered to attend passively to their performance.

The greater part of quality in entertainment arises from the power to reach deeply into the mind of the audience and find a match of resonance. Entertainment that is tuned to the experiences and aspirations of its target audience connects with a crash of cheering, applause, laughter or stunned silence. The teacher who can evoke new qualities of student awareness and understanding releases a flood of productive energy. An entertainment that creates a bond of clear communication enlightens and educates. Herein lies the measure of an entertainment's quality; it reaches into the heart of its audience and strikes a sympathetic chord.

The power to entertain instructs us as to what may be the deepest and most enduring measure of service quality: success in finding the customer in the innermost tabernacle of his or her being. Ultimately, it is probable that we may find the well-spring of service quality in successful entertainment of the customer. To better understand how to do that, we must continue to examine all the factors that have a bearing on service quality.

LE MERIDIEN JAMAICA PEGASUS HOTEL

A Case Study

In October 1998, the largest and premier business hotel in Jamaica, Le Meridien Jamaica Pegasus, became the first hotel in the Caribbean to achieve certification to ISO 9002 Quality Management Standards. This property, referred to as the Pegasus, has 350 rooms and at that time had 400 employees. A unique characteristic of the hotel is the exceptionally long years of service by many senior employees. In 1998 the average length of service to the Pegasus was sixteen years by the first line managers and nine years by the second line managers.

The hotel has four restaurants, two bars, a large ballroom, nine meeting rooms, a gym, two tennis courts, two swimming pools, a jogging track and a shopping arcade. The Pegasus opened in 1972 and has been managed by the same hotel company, Forte Plc. in the United Kingdom, for the last twenty-nine years. During this period, owing to changes within the managing company, the Pegasus underwent two name changes; in 1992 as "Forte Grand" and in 1997 as "Le

Meridien". The second rebranding was more crucial, as to reach Le Meridien standards the hotel had to do considerable upgrading and staff training.

Challenges

In addition to the challenging rebranding process in 1997, the hotel was also faced with a financial problem triggered by an increase of four-star hotel rooms in Kingston by 42% and a decrease of tourist arrivals to Kingston. The occupancy level was around 60% for a few years, but owing to the increase in room capacity in the city, a price war had started. This resulted in the average room rate falling from US$100 to around US$83. Giving in to union demands and agreeing to above-inflation salary increases in the past had resulted in payroll cost (as a percentage of revenue) increasing from 22% in 1995 to 32% in 1997. The bottom line impact of these challenges was that the profitability of the hotel gradually dropped from 22% in 1993 to an alarming 8% in 1997.

Survival Strategies

As a survival strategy the management was compelled to restructure the operations. While there is nothing new to the downsizing phenomenon, the incidence of organization downsizing has increased considerably in recent years. Commentators warn of the potentially damaging effects of downsizing and redundancy programmemes on the survivors and the culture of downsized organizations. The management decided to make 18% of the permanent staff redundant. A bold counterproposal of an unprecedented wage freeze was sent to the union who were demanding a 40% wage increase.

The climate was not appropriate for introducing a quality management system that required high labour moral, motivation and commitment. However, the rebranding to Le Meridien and upgrading of the hotel had been committed with the owners. An official launching of "Le Meridien" (by the prime minister of Jamaica) for the end of 1997 had been confirmed. In this context, the management decided to go ahead with the following during the 1997/1998 financial year:

- Redundancy of fifty-one employees
- Not changing the wage freeze stand in union negotiations
- Frequent frank dialogue with the employees with regards to the financial problems faced by the company and working together in "quality circles" to solve operational problems
- Internal marketing and public relations to convince the employees that quality assurance is the key to solving the financial problems
- Implementing the planned product upgrade
- Completion of the rebranding process by the end of 1997
- Introducing the ISO 9002 Quality Management system as a foundation for the future
- Reviewing the possibility of obtaining four-diamond grading by the American Automobile Association (AAA) and the Green Globe Award within three years

A Vision for the Future

The management decided to pay special attention to training and development of a quality culture at the Pegasus. The vision was that this culture would be the backbone of the future quality standards achievements. It was considered important for this backbone to be strengthened with strategic inputs and quality inputs of the desired standards. At the same time this backbone (quality culture) should be strong enough to hold the present (organizational culture) and the future (vision) together.

Quality Policy

The labour dispute at that time arose as a result of the union's refusal to accept a proposal for a moratorium by the management. In spite of the labour dispute, the management won the support of the employees for the quality vision and 300 employees signed a quality policy, which has been displayed in the lobby of the Pegasus for the last four years. The policy, which was developed with input from the 300 employees, states:

Le Meridien Jamaica Pegasus is committed to being the best up market business hotel in the Caribbean; by consistently delivering Le Meridien quality product and service standards, satisfying guest expectations, whilst building employee morale and achieving profit objectives.

The agreed slogan is, "The Best Is Getting Better".

- All managers meeting (60 minutes) – once a month.
- Managers and union delegates meeting (60 minutes) – once a month

The Le Meridien/quality meeting was a new meeting created to coordinate rebranding and implementing ISO 9002 during the 1997/1998 financial year.

Quality Analysis

Qualitative analysis of all corrective action forms (for every guest complaint) has to be done once a month as part of the ISO 9002 system. Corrective action forms are discussed at the morning briefings and with the use of a log, follow-up action is monitored by the general manager. The commitment from the general manager is essential for the system to work effectively and productively. Guest questionnaires, too, are analysed and distributed to all managers and departments once a month.

Foundation for the Future

The key hotel activities and external factors that had an impact on the hotel's performance from April 1997 to March 1998 are summarized below.

1. People
   - A Le Meridien trainer from Paris trained thirty employees as on-the-job trainers and a series of training programmes for other employees was

conducted by the trainers and management using
Le Meridien manuals as guidelines.

- Twenty-three employees were trained as quality
  auditors to audit the quality of the operations of
  other departments.
- Good relationships were maintained with the
  union and employees in spite of prolonged wage
  negotiations and resulting industrial actions.
- Seven managers underwent brief training/
  exposure at Le Meridien New Orleans in the
  United States.
- Employees of the year were sent to Paris and
  London to join other employees of the year from
  seventy-three Le Meridien hotels on an award
  tour of exposure.
- The executive chef won an award "For Excellence
  in Food Preparation, Delivery and Presentation"
  during 1997, from the *Jamaica Observer*
  newspaper.

2. Product

- Le Meridien image, service and product
  standards were introduced and the hotel was
  rebranded as Le Meridien Jamaica Pegasus on
  October 9, 1997 with the prime minister of
  Jamaica as the guest of honour.
- Various projects were undertaken to implement
  the essential upgrading. These included complete
  refurbishment of two bedroom floors, installation
  of a new chiller, installation of a new fire alarm
  system, painting the building exterior, upgrading
  of tennis courts, upgrading of gardens and public
  areas, etc.
- Food and beverage concepts were changed. The
  coffee shop was upgraded and a section was air-
  conditioned, the gourmet restaurant concept was
  fine-tuned to blend with the historic significance

of Port Royal and the wine club was transformed into a pizza cellar. All menus were changed.

- The ISO 9002 Quality Management system was introduced. The hotel developed a quality policy with input from most of its employees.
- The hotel continued to be the focal point of social activities in Kingston, often being chosen as the location for major banquets. The hotel continued organizing food festivals and other major events, including 101 holiday events during the last 38 days in 1997.
- The hotel won the *Jamaica Observer* newspaper's first ever "Table talk" food award for the "Food Event of the Year" (a six-course gastronomic dinner, served in 1997, for the annual induction ceremony of the prestigious association *La Chaine Des Rotisseurs*).
- The Le Meridien mystery guest programme, with a 1,000point checklist, was introduced by internally arranged monthly mystery guests.

3. Promotions

- Room reservations, banquet bookings, sales, public relations and advertising were combined and a new marketing department was created with the objective of being more aggressive in marketing.
- The start of the twenty-fifth year of operations of the hotel was celebrated with the governor general of Jamaica as the guest of honour.
- Management sales calls to local clients were accelerated.
- The hotel continued to be the focal point during the Jamaica Carnival. A "wet-feete" was held at the hotel as part of the carnival celebrations.
- A major wedding expo was held.
- Le Meridien promoted the hotel in its regional and worldwide publications, sales and advertising campaigns.

- The hotel continued to be in the limelight through planned public relations and publicity.
- Familiarization tours were arranged for Le Meridien vice president–sales, sales director, and reservation agents, to Washington, D.C., New York and London.
- Frequent participation in Le Meridien sales conferences and planned sales trips to North America, South America and Europe.
- Participation in Le Meridien sales promotion activities.

4. Profit

- Towards the end of the financial year, British Airways reduced the allotment of thirty-four rooms per day to six rooms per day as most crew members were provided accommodation in Montego Bay instead of Kingston.
- The number of rooms in Kingston was increased by 32% and the four-star category rooms were increased by 42% within a year. At the same period, the arrivals to Kingston were reduced by 2%. With the supply of rooms in Kingston greatly exceeding the demand, the hotel experienced revenue problems. The hotel undertook major restructuring and re-engineering, affecting fifty-one job positions (18% of the total permanent workforce), which were made redundant. The hotel spent J$11 million as redundancy payments, but was able to recover this within eight months from the savings on labour cost.
- For seven months, the hotel held its position of a wage freeze against the union demand for a 40% wage increase for year one and a 35% increase for year two, for line employees. After two "sit ins" and a strike, the matter was referred to the Industrial Disputes Tribunal and a 7.5%

wage increase for year one and a 7.5% increase for year two was awarded. This is a "landmark" award in the Jamaican hotel industry as salary increases in hotels had been in the range of 30% to 60% for many years prior to this award. As a result of this award, a single-digit wage increase has become the norm in the Jamaican hotel industry today.

Models of Indian Hotels

India's middle class faces a problem that's as typical as it is common: finding hotel accommodation that's safe, clean, comfortable and, most importantly, affordable. More often than not, they have to compromise on one parameter or the other. This is set to change in the coming years following the launch of the indiOne brand. Promoted by Indian Hotels Company, which also operates the Taj Group of Hotels, indiOne is positioned to meet the need for what it terms 'smart basics' accommodation.

Targeted at budget travellers and tourists, indiOne offers an innovative hospitality model where the emphasis is on delivering quality hotel rooms at low cost. The first indiOne property, at Whitefield in Bangalore, opened for business on June 25, 2004, and it has already notched up an occupancy rate of close to 80 per cent. Clearly, this is an idea whose time has come, but the economics of operating and sustaining low-cost, high-quality hotel rooms is not easy.

The indiOne model was arrived at and adopted after extensive qualitative and quantitative research on travel patterns, hotel usage, service needs and the expectations of travellers. The research findings indicated a rich customer base for indiOne, and it included domestic traders, self-employed professionals, pilgrims, backpackers and domestic tourists.

What these people were looking for in their hotel accommodation was pretty much similar: affordability, hygiene and safety on the one hand, and informality, stylishness, warmth and modern amenities on the other.

indiOne is primed to provide all of this and more. Indian Hotels established a wholly owned subsidiary, Roots Corporation, to run the indiOne show, before handing the designing of the project to UK-based architects Young & Gault and the Indian firm, Incubis. Roots is looking to have 1,500 rooms operational under the indiOne umbrella in the next one year, with properties in temple towns and smaller urban centres. The company also has plans to take the brand overseas.

The larger objective behind the launch of indiOne was explained by Ratan Tata, the chairman of the Tata Group, when he unveiled the Bangalore property. "One of the challenges identified [for Indian Hotels] was to innovate and to lead," he said. "This spirit of innovation is evident in the indigenous development of indiOne.

It is a giant step forward for Indian Hotels." Speaking at the same function, Raymond Bickson, the managing director of Indian Hotels, emphasised the business logic powering indiOne. "The dynamics of the entire hospitality industry has changed over the last few years," he said. "A category such as the smart basics hotel has emerged as a compelling business opportunity. We do believe that significant demand exists in the metros and in secondary and tertiary cities across the country."

The Bangalore indiOne has all the standard creature comforts a budget traveller would look for in a hotel room. Besides, it has a cyber cafe, an ATM, safe-deposit boxes, a 24-hour restaurant, a meeting room and a gymnasium. Rooms are air-conditioned and are provided with electronic locks, a 17-inch, flat-screen television and Internet connectivity. There's also a mini-fridge, a tea/coffee maker, hot water, toiletries, same-day laundry services and 24-hour check-in.

The cost for this package is a steal at Rs 900 for a single room for one day and Rs 950 for a double room. The indiOne prototype a reclassification of what constitutes fundamental comforts. We called it 'smart basics' because we have changed the definition of what 'basic' in a hotel connotes. "We have reconfigured a new set of basics in India and value innovation was foremost on our list." With the goodies it has on offer, indiOne is hardly a no-frills hotel in terms of facilities, but this is no five-star extravaganza. Just 25 people run the 101-room Bangalore property.

There is no room service, no porters, and guests have to carry their laundry to the counter. The hotel has 72 single rooms, 20 double rooms (with separate beds for those who travel together) and eight larger-sized rooms. Additionally, there is a special room for the disabled. The launch of indiOne is cause for celebration, not just for Indian Hotels but also for the Indian hospitality industry and for consumers in general.

Hundred years later, the management is walking the global road once again. The world has changed significantly; India has embraced globalisation, along with other countries. The scale of operations in the company has increased. From one hotel in Bombay, the Taj Group has gone on to 53 in India and 12 at various international destinations.

To maintain its status as the very best hotel chain across the globe, the Taj Group is preparing to equip itself to meet the challenges of the next 100 years. The challenge for an organisation that has a tradition of a 100 years lies in how to revisit what you do and improve it so that you don't become stagnant, that you keep up with the trends in the world today.

The Taj is already present in most of the key Indian cities and the scope for domestic growth is limited. Outside India it has hotels in Sri Lanka, Dubai, Oman, Nepal, UK and Maldives. Now the company seeks to expand its vision. The Taj brand is a high profile business for the Group. The group want to become important global players in the Hospitality industry and leverage the Tata brand on an international scale. From India we move westwards to the Indian Ocean (Maldives, Mauritius, Seychelles), then there is Africa and the key gateway cities like London, New York, Shanghai and Beijing." The Australasian and Gulf markets are also opportunity regions.

The success of the Taj Exotica Resort & Spa in Maldives has boosted the confidence of the management, reinforcing that they can compete with the best international brands. For the Taj Group, it is the prototype resort of the future. They have already signed agreements for Mauritius and Seychelles and will soon set up a similar property in Sri Lanka. Four Seasons works on the same model. They have many hotels around the world but own only a few. But the level and standard of service is the same everywhere. That is what we are trying to do too.

Indian Hotels is to be atop the pyramid of the hotel business today and a benchmark in the luxury segment among brands such as Four Seasons and Ritz Carlton. These are aspirational brands and we want to be among them. The attributes for a successful global hotel Group include emphasis on core competence and the basics of the service or the product. You must focus on a few key points that will help reach the goal. The legendary Taj hospitality forms the core competence for Indian Hotels while the key focus areas are renovation, brand building, technology and people training.

Globalisation, like most important matters, begins at home. As the luxury hotels constitute over 70 per cent of the company's profits and attract international guests, the focus is increasingly on revving up the product and the service. In the last five years, some of our competitors have raised the benchmark for luxury. We feel that the quickest way to showcase our hotels is to renovate our luxury palaces.

The Taj Mahal Palace and Towers, Mumbai; Taj Lake Palace, Udaipur; Rambagh Palace, Jaipur; and Falaknuma Palace, Hyderabad; are being refurbished and repositioned in the first step to building the Taj as a global preferred brand and offer global luxury standards and best practises. Luxurious state-of-art spas are replacing the old-fashioned 'health club' concept. A personalised butler service seeks to evoke the lifestyle of the erstwhile Indian maharajas.

The ongoing revolution in cuisine has been accompanied by innovations as well. Though the Group was the first to introduce international cuisine in India, today its Food and Beverages business is competing with free standing, niche restaurants. Thus, new concepts such as contemporary Indian cuisine have been introduced. Internationally recognised chefs and restaurants will also be introduced into the Taj properties.

The challenge in this is that the Taj brand appears on diverse properties from the luxurious Taj Mahal in Delhi to the touristy Taj Gateway in Chiplun. "These products are dissimilar or of very different service standards. The brand does not have a very clear cut luxury connotation. Also, in terms of image, what the brand stands for today may not necessarily be what we want it to stand for in the future.

The issues being debated include bringing about a change in the existing strategic business units, changing the personality of the basic brand (should it be associated with only some hotels?) and the branding of a future international acquisition under the Taj flagship. The marketing and sales teams have been strengthened overseas and are supplemented by PR agencies. Globally the Taj should be reinforced as an international luxury hotel with an Indian soul and touch to it though the degree of Indianness would vary in each country.

Technology is adding new dimensions and increasing efficiency in the Hospitality business. A room is essentially a plain vanilla product. We are trying to add in differentiators through technology. While front-end systems lead to customer satisfaction and product differentiation, the back-end leads to process efficiency and cost savings. In the last three years, the Group has focused on creating the infrastructure and platforms to drive efficiency at both the front and back end.

A Wide Area Network (WAN) now connects all properties allowing for better communications and incorporation of centralised reservation and customer applications. The Taj is also the first chain in the world to have wireless Internet access in most properties. Technology is also helping the Group capture guest trends and preferences to provide more personalised services.

The star in the technology arena for the Taj is the interactive system being rolled out in the heritage wing of the Taj Mumbai. A 42-inch plasma display with surround sound, a personal computer with a wireless keyboard, digital streaming movies or mp3 music gives the guest his private entertainment centre. A cutting edge of product development in hotels, the facility is offered by a select few, like the Dorchester Group.

The Group is leveraging IT resources in the Tata Group to benchmark against the best globally. The Wildfire project, a value driven model catering to what considers a greatly under-served market is based on design and technology. It is utilitarian but contemporary in efficiency and looks. Coming up in areas such as IT parks and industrial towns, the model is scaleable as well as replicable in emerging markets like Afghanistan and Iraq.

While infrastructure and technology can move this industry, it is the people who make it run. The first is upgrading the bench strength. A different mind and skill set is required to go global in the hospitality industry. Positions that call for a global make up have been identified. The global manager's position is crucial to such a set up. Global managers bring with them first-hand experience of global quality and luxury. This leads to cross learning and builds confidence that the Group can compete with the best in the world.

Global managers should be able to understand the nuances of international business, to build a team of people from different cultures and most importantly, to imbibe the Tata culture of compassion and concern. They should be a people's people and culturally sensitive.

High calibre international general managers are being brought into India and underpinned with one or two top Indian managers to enable mentoring. In this way, when the expatriate manager moves to another property, one of the Indian managers can move into his position. About 12 expatriates have already been brought in. An interesting twist to this strategy is to bring in the best global manager who is also an Indian.

Mentoring is followed up with training. The intellectual input involves sending executives for the Harvard general manager programme. Working with an expatriate combined with intellectual training from Harvard can completely change the way people think. The final building block is for young managers to gain overseas work experience when an international property opens up.

In all this, the biggest challenge lies in managing the internal system. The mentoring process or the buddy system was started to counter the anxiety within the company. A Personal Development Plan (PDP) for individuals provides the road map for their career growth. A talent management process is also in place with the help of which every individual has been charted according to potential and ability, and is groomed accordingly. The greatest conviction is that it is not salaries that drive people at work but learning and career goals. With people, products and properties being groomed on this footing, the Taj is ready to take on the globe. Circa 2103, the Taj may well be the jewel not only in India's but the world's crown as well.

Agra Hotel

Agra is famous all around because of its association with the Taj Mahal, one of the Seven Wonders of the World. Agra receives hordes of visitor every year and to cater best service to them, the city offers attractive Agra Hotel Packages.

Jaipur Hotel

The experience of holidaying in Jaipur is like a royal retreat. Hotels in Jaipur are perfect cocoons of luxury and splendor. Various facilities in Jaipur Hotel Packages include the Heritage Hotels, Deluxe Hotels, Star Hotels, Resorts and The Budget Hotels.

Kerala

The mystical land of Kerala, popularly known as Gods Own Country, has everything that tourists desire to explore. The tranquil beaches, long shoreline, greenery and serene backwaters make it the perfect destination.

Uttaranchal

The state with its divine beauty in the form of lush landscapes and meadows with a backdrop of snow-covered mountains offers you a wonderful retreat with Uttaranchal Packages from Indian Holiday.

Himachal

Mighty peaks enveloped in snow, picturesque landscapes, some of the most celebrated religious shrines and thrilling adventure sports - Himachal packages offer you this and much more.

Ladakh

Ladakh Packages offered by Indian Holiday take you to an exotic land full of scenic splendors, gurgling rivers, colourful festivals, impressive monasteries and delightful people.

Kashmir

Visit Kashmir, the crowning glory of India that has no equal in terms of beauty and variety. Kashmir packages offered by Indian Holiday acquaint you to this beautiful state of India.

Nainital

Verdant greens coupled with a salubrious climate make Nainital a popular hill retreat in India. Visit Nainital with Indian Holiday.

South India Hotel

Indian Holiday now offers some delightful South India hotel tour packages for you to choose from. One of the most fascinating tourist destinations in the country, the South Indian region attracts thousands of visitors every year.

Mussoorie

Mussoorie, the majestic "queen of the hills" is an experience to cherish for a lifetime. Right from the delightful colonial charm that pervades the entire hill station to the scenic delights that beckon every visitor

White Water Rafting

Feel the adrenaline coursing through your veins even as you find yourself uttering a whoop of delight as you indulge in a white water rafting exercise in India.

Taj Hotel

Indian Holiday now offers some wonderful Taj Hotel Packages for you to choose from. Enjoy being waited on hand and foot besides reveling in the best of comforts as the Taj hotels take care of all your needs during your India sojourn.

Gulmarg

Experience the tranquil feel and serene environment. Enjoy the velvety meadows and the breathtaking mountains and feel one with the divine on your Gulmarg Dalhousie travel.

Oberoi Hotel

Giving a feeling of home away from home, Oberoi Hotel Packages makes your stay an experience that you will cherish all your life. Indian Holiday makes you experience this luxurious stay with its online booking of Oberoi Hotel Packages.

Index

A

U

Understanding 163

Unemployment Insurance 76

Upstream and downstream sources 147

User 96, 99, 104, 143, 149, 153, 157

Utilization 153

Utilization 40, 44, 99, 153, 240

V

Value bundles 180

Value creation and delivery process 170

Value, customer 3, 4, 5, 6, 8, 11, 15, 16, 17, 19, 23, 25, 26, 27, 28, 30, 83, 94, 105, 106, 110, 125, 131, 139, 148, 157, 158, 165, 170, 172, 176, 179, 182

Value, referral 93, 97

Variability 36, 43

W

Wall Street Journal 1, 3

Wide Area Network 259

Win the Value Revolution 199

X

Xerox and competitor 118

Xerox customers 118

Y

Yellow Pages 104

Young & Gault 256

Youth hostels 51

Z

Zeithaml 105, 106, 203